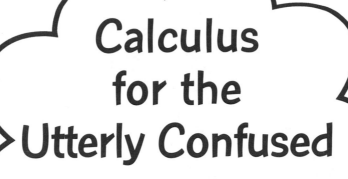

Calculus for the Utterly Confused

Second Edition

Robert Oman
Daniel Oman

McGraw Hill

New York   Chicago   San Francisco   Lisbon   London   Madrid
Mexico City   Milan   New Delhi   San Juan   Seoul
Singapore   Sydney   Toronto

1 2 3 4 5 6 7 8 9 0   DOC/DOC   0 1 2 1 0 9 8 7

ISBN-13: 978-0-07-148158-8
ISBN-10: 0-07-148158-3

This book was set in TimesTen by International Typesetting and Composition and printed on recycled, acid-free paper.

McGraw-Hill books are available at special quantity discounts to use as premiums and sales promotions, or for use in corporate training programs. For more information, please write to the Director of Special Sales, McGraw-Hill Professional, Two Penn Plaza, New York, NY 10121-2298. Or contact your local bookstore.

Cataloging-in-Publication Data is on file with the Library of Congress.

Other books in the Utterly Confused Series include:

Algebra for the Utterly Confused
Astronomy for the Utterly Confused
Beginning French for the Utterly Confused
Beginning Spanish for the Utterly Confused
Chemistry for the Utterly Confused
English Grammar for the Utterly Confused
Financial Aid for the Utterly Confused
Physics for the Utterly Confused
Statistics for the Utterly Confused
Test Taking Strategies and Study Skills for the Utterly Confused

WHAT READERS ARE SAYING

"I wish I had had this book when I needed it most, which was during my pre-med classes. It could have also been a great tool for me in a few medical school courses."

Dr. Kellie Mosley, Recent Medical School Graduate

"Calculus for the Utterly Confused has proven to be a wonderful review enabling me to move forward in application of calculus and advanced topics in mathematics. I found it easy to use and great as a reference for those darker aspects of calculus."

Aaron Landeville, Engineering Student

"I am so thankful for Calculus for the Utterly Confused! I started out clueless but ended with an A!"

Erika Dickstein, Business School Student

"As a non-traditional student one thing I have learned is the importance of material supplementary to texts. Especially in calculus it helps to have a second source, especially one as lucid and fun to read as Calculus for the Utterly Confused. Anyone, whether a math weenie or not, will get something out of this book. With this book, your chances of survival in the calculus jungle are greatly increased."

Brad Baker, Physics Student

To Kelly and Sam

About the Authors

DR ROBERT OMAN received the B.S. degree from Northeastern University and the Sc.M. and Ph.D. degrees from Brown University, all in physics. He has taught mathematics and physics at several colleges and universities including University of Minnesota, Northeastern University, University of South Florida, and University of Tampa. He has also done research for Litton Industries, United Technologies, and NASA, where he developed the theoretical model for the first pressure gauge sent to the moon. He is author of numerous technical articles, books, and how-to-study books, tapes, and videos.

DR DANIEL OMAN received the B.S. degree in physics from Eckerd College. He received both his M.S. degree in physics and his Ph.D. in electrical engineering from the University of South Florida where he taught many *utterly confused* students in classes and one-on-one. He has authored several books and technical articles and has also done research on CO_2 lasers and solar cells. Dan has spent ten years in the semiconductor manufacturing industry with AT&T Bell Labs, Lucent Technologies, and Agere Systems.

Contents

Contents

A Special Message to the Utterly Confused Calculus Student

Our message to the utterly confused calculus student is very simple: You don't have to be confused anymore.

We were once confused calculus students. We aren't confused anymore. We have taught many utterly confused calculus students both in formal class settings and one-on-one. All this experience has taught us what causes confusion in calculus and how to eliminate that confusion. The topics we discuss here are aimed right at the heart of those topics that we know cause the most trouble. Follow us through this book, and you won't be confused anymore either.

Anyone who has taught calculus will tell you that there are two problem areas that prevent students from learning the subject. The first problem is a lack of algebra skills or perhaps a lack of confidence in applying recently learned algebra skills. We attack this problem two ways. One of the largest chapters in this book is the one devoted to a review of the algebra skills you need to be successful in working calculus problems. Don't pass by this chapter. There are insights for even those who consider themselves good at algebra. When we do a problem we take you through the steps, the calculus steps and all those pesky little algebra steps, tricks some might call them. When we give an example it is a complete presentation. Not only do we do the problem completely but also we explain along the way why things are done a certain way.

The second problem of the utterly confused calculus student is the inability to set up the problems. In most problems the calculus is easy, the algebra possibly tedious, but writing the problem in mathematical statements the most difficult step of all. Translating a word problem into a math problem (words to equation) is not easy. We spend time in the problems showing you how to make word sentences into mathematical equations. Where there are patterns to problems we point them out so when you see similar problems, on tests perhaps, you will remember how to do them.

Our message to utterly confused calculus students is simple. You don't have to be confused anymore. We have been there, done that, know what it takes to remove the confusion, and have written it all down for you.

How to Study Calculus

Calculus courses are different from most courses in other disciplines. One big difference is in testing. There is very little writing in a calculus test. There is a lot of mathematical manipulation.

In many disciplines you learn the material by reading and listening and demonstrate mastery of that material by writing about it. In mathematics there is some reading, and some listening, but demonstrating mastery of the material is by doing problems.

For example there is a great deal of reading in a history course, but mastery of the material is demonstrated by writing about history. In your calculus course there should be a lot of problem solving with mastery of the material demonstrated by doing problems.

In your calculus course, practicing working potential problems is essential to success on the tests. Practicing problems, not just reading them but actually writing them down, may be the only way for you to achieve the most modest of success on a calculus test.

To succeed on your calculus tests you need to do three things, PROBLEMS, PROBLEMS, and PROBLEMS. Practice doing problems typical of what you expect on the exam and you will do well on that exam. This book contains explanations of how to do many problems that we have found to be the most confusing to our students. Understanding these problems will help you to understand calculus and do well on the exams.

General Guidelines for Effective Calculus Study

1. If at all possible avoid last minute cramming. It is inefficient.

2. Concentrate your time on your best estimate of those problems that are going to be on the tests.

3. Review your lecture notes regularly, not just before the test.

4. Keep up. Do the homework regularly. Watching your instructor do a problem that you have not even attempted is not efficient.

5. Taking a course is not a spectator event. Try the problems, get confused if that's what it takes, but don't expect to absorb calculus. What you absorb doesn't matter on the test. It is what comes off the end of your pencil that counts.

6. Consider starting an informal study group. Pick people to study with who study and don't whine. When you study with someone agree to stick to the topic and help one other.

Preparing for Tests

1. Expect problems similar to the ones done in class. Practice doing them. Don't just read the solutions.

2. Look for modifications of problems discussed in class.

3. If old tests are available, work the problems.

4. Make sure there are no little mathematical "tricks" that will cause you problems on the test.

Test Taking Strategies

1. Avoid prolonged contact with fellow students just before the test. The nervous tension, frustration, and defeatism expressed by fellow students are not for you.

2. Decide whether to do the problems in order or look over the entire test and do the easiest first. This is a personal preference. Do what works best for you.

3. Know where you are timewise during the test.

4. Do the problems as neatly as you can.

5. Ask yourself if an answer is reasonable. If a return on investment answer is 0.03%, it is probably wrong.

Preface

The purpose of this book is to present basic calculus concepts and show you how to do the problems. The emphasis is on problems with the concepts developed within the context of the problems. In this way the development of the calculus comes about as a means of solving problems. Another advantage of this approach is that performance in a calculus course is measured by your ability to do problems. We emphasize problems.

This book is intended as a supplement in your formal study and application of calculus. It is not intended to be a complete coverage of all the topics you may encounter in your calculus course. We have identified those topics that cause the most confusion among students and have concentrated on those topics. Skill development in translating words to equations and attention to algebraic manipulation are emphasized.

This book is intended for the nonengineering calculus student. Those studying calculus for scientists and engineers may also benefit from this book Concepts are discussed but the main thrust of the book is to show you how to solve applied problems. We have used problems from business, medicine, finance, economics, chemistry, sociology, physics, and health and environmental sciences. All the problems are at a level understandable to those in different disciplines.

This book should also serve as a reference to those already working in the various disciplines where calculus is employed. If you encounter calculus occasionally and need a simple reference that will explain how problems are done this book should be a help to you.

It is the sincere desire of the authors that this book help you to better understand calculus concepts and be able to work the associated problems. We would like to thank the many students who have contributed to this work, many of whom started out utterly confused, by offering suggestions for improvements. Also we would like to thank the people at McGraw-Hill who have confidence in our approach to teaching calculus and support this second edition.

Robert M. Oman
St. Petersburg, Florida

Daniel M. Oman
Singapore

CHAPTER 1

MATHEMATICAL BACKGROUND

Do I Need to Read This Chapter? You should read this chapter if you need to review or you need to learn about

➡ Methods of solving quadratic equations

➡ The binomial expansion

➡ Trigonometric functions—right angle trig and graphs

➡ The various coordinate systems

➡ Basics of logarithms and exponents

➡ Graphing algebraic and trigonometric functions

Some of the topics may be familiar to you, while others may not. Depending on the mathematical level of your course and your mathematical background, some topics may not be of interest to you.

Each topic is covered in sufficient depth to allow you to perform the mathematical manipulations necessary for a particular problem without getting bogged down in lengthy derivations. The explanations are, of necessity, brief. If you are totally unfamiliar with a topic it may be necessary for you to consult an algebra or calculus text for a more thorough explanation.

The most efficient use of this chapter is for you to do a brief review of the topics, spending time on those sections that are unfamiliar to you and that you know you will need in your course, and then refer to specific topics as they are encountered in the solution to problems. Even if you are familiar with a topic review might "fill in the gaps" or give you a better insight into certain mathematical operations. With this reference you should be able to perform all the algebraic operations necessary to complete the problems in your calculus course.

1-1   Solving Equations

The simplest equations to solve are the linear equations of the form $ax + b = 0$, which have as their solution $x = -b/a$. The next most complicated equations are the quadratics. The simplest quadratic is the type that can be solved by taking square roots directly.

Example 1-1   Solve for x: $4x^2 = 36$.

Solution: Divide by 4, and then take the square root of both sides.

$$\frac{4x^2}{4} = \frac{36}{4} \;\Rightarrow\; x^2 = 9 \;\Rightarrow\; x = \pm 3$$

Both plus and minus values are legitimate solutions. The reality of the problem producing the equation may dictate that one of the solutions be discarded.

The next complication in quadratic equations is the factorable equation.

Example 1-2   Solve $x^2 - x - 6 = 0$ by factoring.

Solution: $x^2 - x - 6 = 0 \;\Rightarrow\; (x - 3)(x + 2) = 0$ The solutions, the values of x that make each parenthesis equal to zero, and satisfy the factored equation, are $x = 3$ and $x = -2$.

If the quadratic cannot be solved by factoring, the most convenient solution is by quadratic formula, a general formula for solution of any quadratic equation in the form $ax^2 + bx + c = 0$. The solution according to the quadratic formula is

$$x = \frac{-b \pm \sqrt{b^2 - 4ac}}{2a}$$

Problems in your course should rarely produce square roots of negative numbers. If your solution to a quadratic produces any square roots of negative numbers, you are probably doing something wrong in the problem.

Example 1-3 Solve $x^2 - 5x + 3 = 0$ by using the quadratic formula.

Solution: Substitute the constants into the formula and perform the operations. Writing $ax^2 + bx + c = 0$ above the equation you are solving helps in identifying the constants and keeping track of the algebraic signs.

$$ax^2 + bx + c = 0$$
$$x^2 - 5x + 3 = 0$$

$$x = \frac{-b \pm \sqrt{b^2 - 4ac}}{2a} = \frac{+5 \pm \sqrt{25 - 4(1)(3)}}{2(1)} = \frac{5 \pm \sqrt{13}}{2} = 4.30, 0.70$$

The quadratic formula comes from a generalized solution to quadratics known as "completing the square." Completing the square is rarely used in solving quadratics. The formula is much easier. It is, however, used in certain calculus problems, so an explanation of the technique is appropriate. A completing the square approach is also used in graphing certain functions.

The basic procedure for solving by completing the square is to make the equation a perfect square, much as was done with the simple example, $4x^2 = 36$. Work with the x^2 and x coefficients so as to make a perfect square of both sides of the equation and then solve by direct square root. This is best seen by example.

Look first at the equation $x^2 + 6x + 5 = 0$, which can be factored and has solutions of -5 and -1, to see how completing the square produces these solutions.

Example 1-4 Solve $x^2 + 6x + 5 = 0$ by completing the square.

Solution: The equation can be made into a perfect square by adding 4 to both sides of the equation to read $x^2 + 6x + 9 = 4$ or $(x + 3)^2 = 4$, which, upon direct square root, yields $x + 3 = \pm 2$, producing solutions -5 and -1.

As you can imagine the right combination of coefficients of x^2 and x can make the problem awkward. Most calculus problems involving completing the square are not especially difficult. The general procedure for completing the square is the following:

- If necessary, divide to make the coefficient of the x^2 term equal to 1.
- Move the constant term to the right side of the equation.
- Take 1/2 of the x coefficient, square it, and add to both sides of the equation. This makes the left side a perfect square and the right side a number.
- Write the left side as a perfect square and take the square root of both sides for the solution.

Example 1-5   Solve $x^2 + 4x + 1 = 0$ by completing the square.

Solution: Move the 1, the constant term, to the right side: $x^2 + 4x = -1$. Add 1/2 of 4 (the coefficient of x) squared to both sides: $x^2 + 4x + 4 = 4 - 1$. The left side is a perfect square and the right side a number: $(x + 2)^2 = 3$. Take square roots for the solutions: $x + 2 = \pm\sqrt{3}$ or $x = -2 + \sqrt{3}, \ -2 - \sqrt{3}$.

Certain cubic equations such as $x^3 = 8$ can be solved directly producing the single answer $x = 2$. Cubic equations with quadratic (x^2) and linear (x) terms can be solved by factoring (if possible) or approximated using graphical techniques. Calculus will allow you to apply graphical techniques to solving cubics.

1-2   Binomial Expansions

Squaring $(a + b)$ is done so often that most would immediately write $a^2 + 2ab + b^2$. Cubing $(a + b)$ is not so familiar but easily accomplished by multiplying $(a^2 + 2ab + b^2)$ by $(a + b)$ to obtain $a^3 + 3a^2b + 3ab^2 + b^3$.

There is a simple procedure for finding the nth power of $(a + b)$. Envision a string of $(a + b)$s multiplied together $(a + b)^n$. Notice that the first term has coefficient 1 with a raised to the nth power, and the last term has coefficient 1 with b raised to the nth power. The terms in between contain a to progressively decreasing powers, $n, n - 1, n - 2, \ldots$, and b to progressively increasing powers, $0, 1, 2, \ldots$ The coefficients can be obtained from an array of numbers or more conveniently from the binomial expansion or binomial theorem

$$(a + b)^n = \frac{a^n}{0!} + \frac{na^{n-1}b}{1!} + \frac{n(n - 1)a^{n-2}b^2}{2!} + \cdots$$

The factorial notation may be new to you. The definitions are

$$0! = 1, \quad 1! = 1, \quad 2! = 2 \cdot 1, 3! = 3 \cdot 2 \cdot 1, \quad \text{etc.}$$

Use of the binomial expansion to verify $(a + b)^3$ is one of the suggested problems.

1-3 Trigonometry

The trigonometric relations can be defined in terms of right angle trigonometry or through their functions. The basic trigonometric relations, as they relate to right triangles, are shown in the box below.

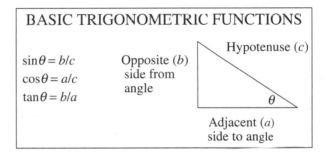

Graphs of the trigonometric relations are shown in Fig. 1-1.

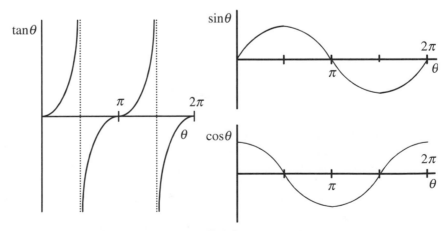

Fig. 1-1

The tangent function is also defined in terms of sine and cosine: $\tan\theta = \sin\theta/\cos\theta$.

Angles are measured in radians and degrees. $\theta = s/r$ or $s = r\theta$ Radian measure is a pure number, the ratio of arc length to radius to produce the desired angle. Figure 1-2 shows the relationship of arc length to radius to define the angle.

Fig. 1-2

The relation between radians and degrees is 2π rad $= 360°$.

Example 1-6   Convert $\pi/6$ and 0.36 rad to degrees and 270° to radians.

Solution: $\dfrac{\pi}{6}$ rad $\dfrac{360°}{2\pi\,\text{rad}} = 30°$,   0.36 rad $\dfrac{360°}{2\pi\,\text{rad}} = 20.6°$,

$270° \dfrac{2\pi\,\text{rad}}{360°} = \dfrac{3\pi}{2}$ rad $= 4.7\,\text{rad}$

There are a large number of trigonometric identities that can be derived using geometry and algebra. Several of the more common are in the following box:

TRIGONOMETRIC IDENTITIES

$a^2 + b^2 = c^2$ $\sin^2\theta + \cos^2\theta = 1$

$\sin\theta = \cos(90° - \theta)$ $\cos\theta = \sin(90° - \theta)$

$\sin(\alpha \pm \beta) = \sin\alpha\cos\beta \pm \cos\alpha\sin\beta$ $\tan\theta = \dfrac{1}{\tan(90° - \theta)}$

$\cos(\alpha \pm \beta) = \cos\alpha\cos\beta \mp \sin\alpha\sin\beta$

$\tan(\alpha \pm \beta) = \dfrac{\tan\alpha \pm \tan\beta}{1 \mp \tan\alpha\,\tan\beta}$

1-4   Coordinate Systems

The standard two-dimensional coordinate system works well for most calculus problems. In working problems in two dimensions do not hesitate to arrange the coordinate system for your convenience. The *x*-coordinate does not have to be horizontal and increasing to the right. It is best, however, to maintain the *x-y* orientation. With the fingers of the right hand pointed in the direction of *x* they should naturally curl in the direction of *y*. Positions in the standard right angle coordinate system are given with two numbers. In a polar coordinate system positions are given by a number and an angle. In Fig. 1-3 it is clear that any point (*x*, *y*) can also be specified by (*r*, θ).

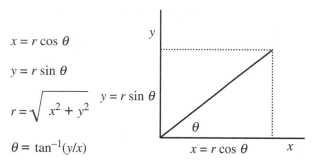

$x = r \cos \theta$

$y = r \sin \theta$

$r = \sqrt{x^2 + y^2}$

$\theta = \tan^{-1}(y/x)$

Fig. 1-3

Rather than moving distances in mutually perpendicular directions, the r and θ locate points by moving a distance r from the origin along what would be the $+x$ direction, then rotating counterclockwise through an angle θ. The relationship between rectangular and polar coordinates is also shown in Fig. 1-3.

Example 1-7 Find the polar coordinates for the point $(3, 4)$.

Solution: $r = \sqrt{3^2 + 4^2} = 5$ and $\theta = \tan^{-1}(4/3) = 53°$

Quick Tip

Be sure that you understand how to calculate $\theta = \tan^{-1}(4/3) = 53°$ on your calculator. This is not $1/\tan(4/3)$. This is the inverse tangent. Instead of the ratio of two sides of a right triangle (the regular tangent function), the inverse tangent does the opposite: it calculates the angle from a number, the ratio of the two sides of the triangle. On most calculators you need to hit a 2nd function key or "inv" key to perform this "inverse" operation.

Example 1-8 Find the rectangular points for $(3, 120°)$.

Solution: $x = 3 \cos 120° = -1.5$ and $y = 3 \sin 120° = 2.6$

As a check, you can verify that $(-1.5)^2 + 2.6^2 = 3^2$.

Three-dimensional coordinate systems are usually right handed. In Fig. 1-4 imagine your right hand positioned with fingers extended in the $+x$ direction closing naturally so that your fingers rotate into the direction of the $+y$ axis while your thumb points in the direction of the $+z$ axis. It is this rotation of

x into y to produce z with the right hand that specifies a right-handed coordinate system. Points in the three-dimensional system are specified with three numbers (x, y, z).

For certain types of problems, locating a point in space is more convenient with a cylindrical coordinate system, as shown in Fig. 1-5. Notice that this is also a right-handed coordinate system with the central axis of the cylinder as the z-axis.

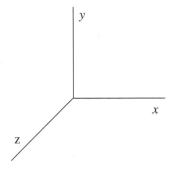

Fig. 1-4

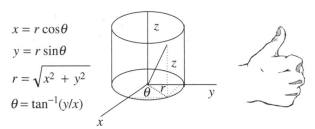

$$x = r\cos\theta$$
$$y = r\sin\theta$$
$$r = \sqrt{x^2 + y^2}$$
$$\theta = \tan^{-1}(y/x)$$

Fig. 1-5

A point is located by specifying a radius measured out from the origin in the $+x$ direction, an angle in the x-y plane measured from the x-axis, and a height above the x-y plane. Thus the coordinates in the cylindrical system are (r, θ, z). The relation of these coordinates to x, y, z is given in Fig. 1-5.

1-5　Logarithms and Exponents

Logarithms and exponents are used to describe several physical phenomena. The exponential function $y = a^x$ is a unique one with the general shape shown in Fig. 1-6.

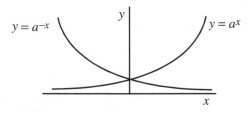

Fig. 1-6

This exponential equation $y = a^x$ cannot be solved for x using normal algebraic techniques. The solution to $y = a^x$ is one of the definitions of the logarithmic function:

$$y = a^x \Rightarrow x = \log_a y$$

The language of exponents and logarithms is much the same. In exponential functions we say "a is the base raised to the power x." In logarithm functions we say "x is the logarithm to the base a of y." The laws for the manipulation of exponents and logarithms are similar. The manipulative rules for exponents and logarithms are summarized in the box below.

The term "log" is usually used to mean logarithms to the base 10, while "ln" is used to mean logarithms to the base e. The terms "natural" (for base e) and "common" (for base 10) are frequently used.

LAWS OF EXPONENTS AND LOGARITHMS

$$(a^x)^y = a^{xy} \quad y \log_a x = \log_a x^y$$

$$a^x a^y = a^{x+y} \quad \log_a x + \log_a y = \log_a xy$$

$$\frac{a^x}{a^y} = a^{x-y} \quad \log_a x - \log_a y = \log_a \frac{x}{y}$$

Example 1-9   Convert the exponential statement $100 = 10^2$ to a logarithmic statement.

Solution: $y = a^x$ is the same statement as $x = \log_a y$, so $100 = 10^2$ is $2 = \log_{10} 100$.

Example 1-10   Convert the exponential statement $e^2 = 7.4$ to a (natural) logarithmic statement.

Solution: $e^2 = 7.4$ so $\ln 7.4 = 2$

Example 1-11   Convert $\log 2 = 0.301$ to an exponential statement.

Solution: $10^{0.301} = 2$

Example 1-12   Find $\log(2.1)(4.3)^{1.6}$.

Solution: On your hand calculator raise 4.3 to the 1.6 power and multiply this result by 2.1. Now take the log to obtain 1.34.

Second Solution: Applying the laws for the manipulation of logarithms write:

$$\log(2.1)(4.3)^{1.6} = \log 2.1 + \log 4.3^{1.6} = \log 2.1 + 1.6\log 4.3 = 0.32 + 1.01 = 1.33$$

(Note the round-off error in this second solution.) This second solution is rarely used for numbers. It is, however, used in solving equations.

Example 1-13   Solve $4 = \ln 2x$.

Solution: Apply a manipulative rule for logarithms: $4 = \ln 2 + \ln x$ or $3.31 = \ln x$.

Now switch to exponentials: $x = e^{3.31} = 27.4$

Quick Tip

A very convenient phrase to remember in working with logarithms is "a logarithm is an exponent." If the logarithm of something is a number or an expression, then that number or expression is the exponent of the base of the logarithm.

1-6   Functions and Graphs

Functions can be viewed as a series of mathematical orders. The typical function is written starting with y, or $f(x)$, read as "f of x," short for function of x. The mathematical function y or $f(x) = x^2 + 2x + 1$ is a series of orders or operations to be performed on an as yet to be specified value of x. This set of orders is: square x, add 2 times x, and add 1. The operations specified in the function can be performed on individual values of x or graphed to show a continuous "function." It is the graphing that is most encountered in calculus. We'll look at a variety of algebraic functions eventually leading into the concept of the limit.

Example 1-14   Perform the functions $f(x) = x^3 - 3x + 7$ on the number 2 or find $f(2)$.

Solution: Performing the operations on the specified function

$$f(2) = 2^3 - 3(2) + 7 = 8 - 6 + 7 = 9$$

In visualizing problems it is very helpful to know what certain functions look like. You should review the functions described in this section until you can look at a function and picture "in your mind's eye" what it looks like. This skill will prove valuable to you as you progress through your calculus course.

Linear

The linear algebraic function (see Fig. 1-7) is $y = mx + b$, where m is the slope of the straight line and b is the intercept, the point where the line crosses the y-axis. This is not the only form for the linear function, but it is the one that is used in graphing and is the one most easily visualized.

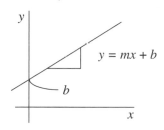

Fig. 1-7

Example 1-15   Graph the function $y = 2x - 3$.

Solution: This is a straight line, and it is in the correct form for graphing. Because the slope is positive, the curve rises with increasing x. The coefficient 2 tells you that the curve is steeper than a slope 1, (which has a 45° angle). The constant -3 is the intercept, the point where the line crosses the y-axis. (See Fig. 1-8.)

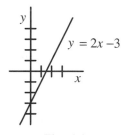

Fig. 1-8

You should go through this little visualization exercise with every function you graph. Knowing the general shape of the curve makes graphing much easier. With a little experience you should look at this function and immediately visualize that (1) it is a straight line (first power), (2) it has a positive slope greater than 1 so it is a rather steep line rising to the right, and (3) the constant term means that the line crosses the y-axis at -3.

Knowing generally what the line looks like, place the first (easiest) point at $x = 0$, $y = -3$. Again knowing that the line rises to the right, pick $x = 2$, $y = 1$, and as a check $x = 3$, $y = 3$.

If you are not familiar with visualizing the function before you start calculating points, graph a few straight lines, but go through the exercise outlined above before you place any points on the graph.

Quadratics

The next most complicated function is the quadratic (see Fig. 1-9), and the simplest quadratic is $y = x^2$, a curve of increasing slope, symmetric about the y-axis (y has the same value for $x = +$ or -1, $+$ or -2, etc.). This symmetry property is very useful in graphing. Quadratics are also called parabolas. Adding a constant to obtain $y = x^2 + c$ serves to move the curve up or down the y-axis in the same way the constant term moves the straight line up and down the y-axis.

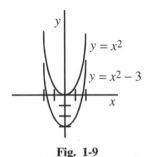

Fig. 1-9

Example 1-16   Graph $y = x^2 - 3$.

Solution: First note that the curve is a parabola with the symmetry attendant to parabolas and it is moved down on the y-axis by the -3. The point $x = 0$, $y = -3$ is the key point, being the apex or lowest point for the curve, and the defining point for the symmetry line, which is the y-axis. Now, knowing the general shape of the curve add the point $x = \pm 2$, $y = 1$. This is sufficient information to construct the graph as shown in Fig. 1-9. Further points can be added if necessary.

Adding a constant a in front of the x^2 either sharpens ($a > 1$) or flattens ($a < 1$) the graph. A negative a value causes the curve to open down.

Example 1-17   Graph $y = 0.5x^2 + 1$.

Solution: Looking at the function, note that it is a parabola (x^2 term), it is flatter than normal (0.5 coefficient), it opens up (positive coefficient of the x^2 term), and it is moved up the axis one unit. Now put in some numbers: $x = 0$, $y = 1$ is the apex, and the y-axis is the symmetry line. Add the points $x = \pm 2$, $y = 3$ and sketch the graph (Fig. 1-10).

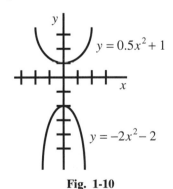

Fig. 1-10

Example 1-18   Graph $y = -2x^2 - 2$.

Solution: Look at the function and verify the following statement. This is a parabola that opens down, is sharper than normal, and is displaced two units in the negative direction. Put in the two points $x = 0$ and $x = \pm 1$ and verify the graph shown in Fig. 1-10.

Adding a linear term, a constant times x, so that the function has the form $y = ax^2 + bx + c$ produces the most complicated quadratic. The addition of this constant term moves the curve both up and down and sideways. If the quadratic function is factorable then the places where it crosses the x-axis are obtained directly from the factored form.

Example 1-19 Graph the function $y = f(x) = x^2 + 2x - 8$.

Solution: This is a parabola that opens up, and is displaced up or down and sideways. This quadratic is factorable to $y = (x + 4)(x - 2)$. The values $x = 2$ and $x = -4$ make $y = 0$, so these are the points where the curve crosses the x-axis. Place these points on the graph.

Now here is where the symmetry property of parabolas is used. Because of the symmetry, the parabola must be symmetric about a line halfway between $x = 2$ and $x = -4$, or about the line $x = -1$. The apex of the parabola is on this $x = -1$ line so substitute to find the appropriate value of y: $f(-1) = (-1+4)$ $(-1-2) = -9$. These three points are sufficient to sketch the curve (see Fig. 1-11).

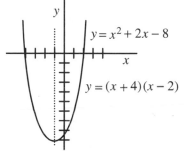

Fig. 1-11

Before moving on to the graphing of quadratics that are not factorable there is one other quadratic that is rather simple yet it illustrates the method necessary for rapid graphing of nonfactorable quadratics.

Example 1-20 Graph $y = x^2 + 4x + 4$.

Solution: Notice in Fig. 1-12 that the right side of this equation is a perfect square and the equation can be written as $y = (x + 2)^2$. The apex of the curve is at $x = -2$, and any variation of x from -2 makes y positive. And the parabola is symmetric about the line $x = -2$. If $x = -1$ or $x = -3$, $y = 1$. If $x = 0$ or $x = -4$, $y = 4$. This is sufficient information to sketch the curve. Notice, however, in the second solution an even easier means for graphing the function.

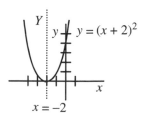

Fig. 1-12

Second Solution: The curve can be written in the form $y = X^2$ if X is defined as $X = x + 2$. At $x = -2$, $X = 0$ and the line $x = -2$ effectively defines a new axis. Call it the Y-axis. This is the axis of symmetry determined in the previous solution. Drawing in the new axis allows graphing of the simple equation $y = X^2$ about this new axis.

Pattern

The previous problem and the following problem illustrate the progression of perfect square to completing the square approaches to graphing quadratics. This is a valuable time-saving method of graphing.

Example 1-21   Graph $y = x^2 - 6x + 11$.

Solution: Based on experience with the previous problem subtract 2 from both sides to at least get the right side a perfect square: $y - 2 = x^2 - 6x + 9 = (x - 3)^2$. This form of the equation suggests the definitions $Y = y - 2$ and $X = x - 3$, so that the equation reads $Y = X^2$. This is a parabola of standard shape on the new coordinate system with origin at (3,2). The new coordinate axes are the lines $x = 3$ and $y = 2$. This rather formidable looking function can now be drawn quite easily with the new coordinate axes. (See Fig. 1-13.)

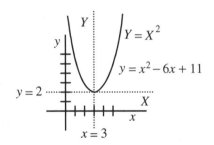

Fig. 1-13

The key step in getting started on Example 1-21 is recognizing that subtracting 2 from both sides makes a perfect square on the right. This step is not always obvious, so we need a method of converting the right-hand side into a perfect square. This method is a variation of the "completing the square" technique for solving quadratic equations. If you are not very familiar with completing the square (this should include nearly everyone) go back in this chapter and review the process before going on. Now that you have "completing the square" clearly in your mind we'll graph a nonfactorable quadratic with a procedure that always works.

Example 1-22   Graph $y = x^2 + 4x + 7$.

Solution: 1. Move the constant to the left side of the equation: $y - 7 = x^2 + 4x$. Next, determine what will make the right-hand side a perfect square. In this case

+4 makes a perfect square on the right so add this to both sides: $y - 3 = x^2 + 4x + 4$ or $y - 3 = (x + 2)^2$.

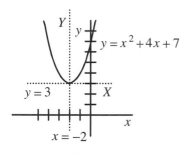

2. Now, make the shift in axes with the definitions $Y = y - 3$ and $X = x + 2$. The origin of the "new" coordinate axes is $(-2,3)$. Determining the origin from these defining equations helps to prevent scrambling the $(-2,3)$ and getting the origin in the wrong place. The values $(-2,3)$ make X and Y zero and this is the apex of the curve $Y = X^2$ on the new coordinate axes.

Fig. 1-14

3. Graph the curve as shown in Fig. 1-14.

Higher Power Curves

The graphing of cubic and higher power curves requires techniques you will learn in your calculus course. There are, however, some features of higher power curves that can be learned from an "algebraic" look at the curves.

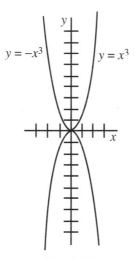

The simple curves for $y = x^3$ and $y = -x^3$ are shown in Fig. 1-15. Adding a constant term to either of these curves serves to move them up or down on the y-axis the same as it does for a quadratic or straight line. Cubics plus a constant are relatively easy to sketch. Adding a quadratic or linear term adds complications that are almost always easiest handled by learning the calculus necessary to help you graph the curve. If a curve contains an x^3 term, this term will eventually predominate for sufficiently large x.

Fig. 1-15

Operationally, this means that if you have an expression $y = x^3 + ()x^2 + ()x + ()$, while there may be considerable gyration of the curve near the origin, for large (positive or negative) x the curve will eventually take the shape shown in Fig. 1-15.

The same is true for other higher power curves. The curve $y = x^4$ is similar in shape to $y = x^2$, it just rises more rapidly. The addition of other (lower than 4) power terms again may add some interesting twists to the curve but for large x it will eventually rise sharply.

1-7   Conics

The next general category of curves is called *conics*, because they have shapes generated by passing a plane through a cone. They contain x and y terms to the second power. The simplest of these curves is generated with x^2 and y^2 equal to a constant. More complicated curves have positive coefficients for these terms, and the most complicated conics have positive and negative coefficients.

Circles

Circles are functions in the form $x^2 + y^2 = $ const. with the constant written in what turns out to be a convenient form, $x^2 + y^2 = r^2$. The curve $x^2 + y^2 = r^2$ is composed of a collection of points in the x-y plane whose squares equal r^2. Look at Fig. 1-16 and note that for each (x,y) point that satisfies the equation, a right triangle can be constructed with sides x, y, and r, and the Pythagorean theorem defines the relationship $x^2 + y^2 = r^2$. A circle is a collection of points, equal distance from a point called the *center*.

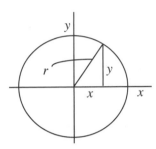

Fig. 1-16

Example 1-23   Graph $x^2 + y^2 = 9$.

Solution: Look at the function and recognize that it is a circle. It has radius 3 and it is centered about the origin. At $x = 0$, $y = \pm 3$, and at $y = 0$, $x = \pm 3$. Now draw the circle (Fig. 1-17). Note that someone may try to confuse you by writing this function as $y^2 = 9 - x^2$. Don't let them.

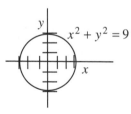

Fig. 1-17

Example 1-24   Graph $x^2 - 6x + 9 + y^2 = 16$.

Solution: At first glance it looks as though a page is missing between Examples 1-23 and 1-24. But if you make the identification that $x^2 - 6x + 9$ is the perfect square of $(x - 3)$ then the equation reads $X^2 + Y^2 = 16$ if $X = x - 3$ and $Y = y$. This is the identification that worked so well for parabolas. In the new coordinate system with origin at $(3,0)$ this curve is a circle of radius 4, centered on the point $(3,0)$ (Fig. 1-18). Set up the new coordinate system and graph the circle. At $X = 0$, $Y = \pm 4$, and at $Y = 0$, $X = \pm 4$.

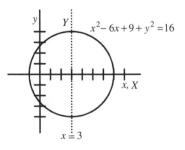

Fig. 1-18

If the function were written $y^2 = 6x - x^2 + 7$ it would not have been quite so easy to recognize the curve. Looking at this latter rearrangement, the clue that this is a circle is that the x^2 and y^2 terms are both positive when they are together on the same side of the equation. No matter how scrambled the terms are, if you can recognize that the curve is a circle you can separate out the terms and make some sense out of them by making perfect squares. This next problem will give you an example that is about as complicated as you will encounter.

Example 1-25   Graph $x^2 + 6x + y^2 + 2y = 10$.

Solution: Notice that the x and y terms are at least grouped together and further that the constant has been moved to the right side of the equation. This is similar to the first step in solving an equation by completing the square. Now with the equation written in this form write the perfect squares that satisfy the x^2 and x terms and the y^2 and y terms adding the appropriate constants to the right side.

$$x^2 + 6x + y^2 + 2y = 10$$

$$(x + 3)^2 + (y + 1)^2 = 10 + 9 + 1$$

$$(x + 3)^2 + (y + 1)^2 = 20$$

Make the identification

$X = x + 3$ and $Y = y + 1$ so $X^2 + Y^2 = 20$

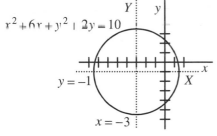

This is a circle of radius $\sqrt{20}$ centered about the point $(-3,-1)$ (Fig. 1-19). A rather formidable function is not so difficult when viewed properly.

Fig. 1-19

Circles can at first be very confusing. If the x^2 and y^2 coefficients can be made equal to 1 and they are positive, then you are dealing with a circle. Knowing the curve is a circle is a long way toward drawing it correctly.

Ellipses

Ellipses have x^2 and y^2 terms with positive but different coefficients. The two forms for the equation of an ellipse are

$$ax^2 + by^2 = c^2 \quad \text{or} \quad \frac{x^2}{a^2} + \frac{y^2}{b^2} = 1$$

Each form has its advantages with the latter form being the more convenient for graphing.

Example 1-26 Graph $4x^2 + 9y^2 = 36$.

Solution: This is an ellipse because the x and y terms are squared and have different positive coefficients. The different coefficients indicate a stretching or compression of the curve in the x or y direction. It is not necessary to know the direction. That comes out of the graphing technique. Rewrite the equation into a more convenient form for graphing by dividing by 36.

$$\frac{x^2}{9} + \frac{y^2}{4} = 1$$

Now in this form set first $x = 0$, so $y = \pm2$, and then $y = 0$, so $x = \pm3$. With these points and the knowledge that it is a circle compressed in one direction, sketch the curve (Fig. 1-20).

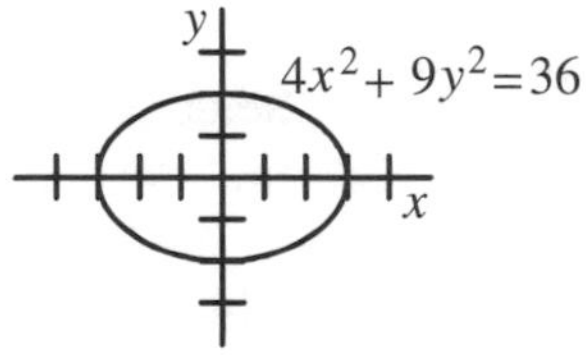

Fig. 1-20

Example 1-27 Graph $\dfrac{(x + 1)^2}{16} + \dfrac{(y - 4)^2}{36} = 1$.

Solution: The problem is presented in this somewhat artificial form to illustrate the axis shifting used so effectively in the graphing of parabolas and circles.

Based on this experience immediately write

$$\frac{X^2}{16} + \frac{Y^2}{36} = 1$$

with the definitions $X = x + 1, Y = y - 4$.

The origin of the new coordinate system is at $(-1,4)$, and in this new coordinate system when $X = 0$, $Y = \pm6$ and when $Y = 0$, $X = \pm4$.

Sketch the curve (Fig. 1-21).

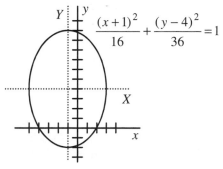

Fig. 1-21

Example 1-28 Graph $x^2 + 4x + 9y^2 - 18y = -4$.

Solution: The different positive coefficients of the x^2 and y^2 terms tell us this is an ellipse. The linear terms in x and y tell us it is displaced off the x-y axis.

Graphing this curve is going to require a completing the square approach with considerable attention to detail.

First, write $x^2 + 4x + 9(y^2 - 2y) = -4$.

Now do the completing the square exercise, being very careful of the 9 outside the parentheses: $(x + 2)^2 + 9(y - 1)^2 = -4 + 4 + 9 = 9$

$$(x^2 + 4x + 4) + 9(y^2 - 2y + 1) = -4 + 4 + 9 = 9$$

Now divide to reach $\dfrac{(x + 2)^2}{9} + \dfrac{(y - 1)^2}{1} = 1$

Define $X = x + 2$ and $Y = y - 1$ to achieve

$$\frac{X^2}{9} + \frac{Y^2}{1} = 1$$

Graph on the new coordinate system: when $X = 0$, $Y = \pm 1$, and when $Y = 0$, $X = \pm 3$.

Alternate Solution: An alternative to graphing in the new coordinate system is to go back to the original coordinate system. When $X = \pm 3$, substitute and write $x + 2 = \pm 3$ or $x = -2 \pm 3$, and when $Y = \pm 1$, substitute and write $y - 1 = \pm 1$ or $y = 1 \pm 1$. Either way gives the same points on the graph (Fig. 1-22).

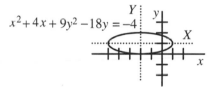

$x^2 + 4x + 9y^2 - 18y = -4$

Fig. 1-22

Hyperbolas

Ellipses are different from circles because of numerical coefficients for the x^2 and y^2 terms. Hyperbolas are different from ellipses and circles because one of the coefficients of these x^2 and y^2 terms is negative. This makes the analysis somewhat more complicated. Hyperbolas are written in one of two forms, both of which are sometimes needed in the graphing.

$$\pm ax^2 \mp by^2 = c^2 \quad \text{or} \quad \pm \frac{x^2}{a^2} \mp \frac{y^2}{b^2} = 1$$

Example 1-29 Graph $-4x^2 + 25y^2 = 100$.

Solution: The form of the equation tells us this is a hyperbola. Now proceed as if this were a circle or ellipse: If $x = 0$, $y = \pm 2$, and if $y = 0$, there are no real values of x. If the curve goes through the points $(0,2)$ and $(0,-2)$ and does not exist along the line $y = 0$, then the curve must have two separate parts! Rearrange the equation to $4x^2 = 25y^2 - 100$ and note immediately that for real values of x, y has to be greater than 2 or less than -2. The curve does not exist in the region bounded by the lines $y = 2$ and $y = -2$.

At this point in the analysis we have two points and a region where the curve does not exist. Further analysis requires a departure from the usual techniques applied to conics. Rewrite the equation again, but this time in the form $y = ...$

$$25y^2 = 4x^2 + 100$$

$$y^2 = \frac{4x^2}{25} + 4$$

$$y = \pm\sqrt{4x^2/25 + 4}$$

How this helps in graphing is that for large values of x, the function begins to look like a straight line, $y \approx \pm(2/5)x$ (for large x the $+4$ is small compared to $4x^2/25$). Use these two straight lines, one of slope $(2/5)$ and the other of slope $-(2/5)$, as guides in drawing the curve. With the points $(0,2)$ and $(0,-2)$ and these lines as guides, the curve can be sketched (Fig. 1-23). In the language of mathematics these straight lines are asymptotes or asymptote lines. *Asymptotes* are lines the curve approaches but does not touch.

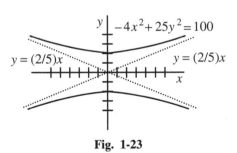

Fig. 1-23

Now that you know the general shape of hyperbolas, we can look at some hyperbolas that are not symmetric about the origin. The next problem is somewhat artificial, but it is instructive and illustrates a situation that comes up in the graphing of hyperbolas.

Example 1-30 Graph $\dfrac{(x-1)^2}{9} - \dfrac{(y-3)^2}{4} = 1$.

Solution: This function is in a convenient form for graphing, especially if we make the identification $X = x - 1$ and $Y = y - 3$. This hyperbola is displaced up and down and sideways to the new coordinate system with origin at $(1,3)$. In this new coordinate system at $X = 0$, Y does not have any real values. At $Y = 0$, $X = \pm 3$. Place these points on the graph.

The asymptote lines are most easily drawn in the new coordinate system.

The transformed function is

$$\frac{(x-1)^2}{9} - \frac{(y-3)^2}{4} = 1$$

$$\frac{X^2}{9} - \frac{Y^2}{4} = 1$$

$Y^2 = (4/9)X^2 - 4$ and for large values of X, $Y \approx \pm(2/3)X$ $Y \approx \pm(2/3)X$

$$9Y^2 = 4X^2 - 36$$

Straight lines of slope $+(2/3)$ and $-(2/3)$ are drawn in the new coordinate system. With the two points and these asymptote lines the curve can be sketched.

In Fig. 1-24 you will see a rectangle. This is used by some as a convenient construct for drawing the asymptote lines and finding the critical points of the curve. Two sides of the rectangle intersect the X-axis at the points where the curve crosses this axis and the diagonals of the rectangle have slopes $\pm (2/3)$.

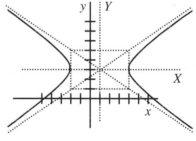

Fig. 1-24

Example 1-31 Graph $9x^2 - 4y^2 - 54x - 32y = 19$.

Solution: This is a hyperbola, and the presence of the linear terms indicates it is moved up and down and sideways. Graphing requires a completing the square approach. Follow the completing the square approach through the equations below. Watch the multiplication of the parentheses very carefully.

$$9x^2 - 4y^2 - 54x - 32y = 19$$
$$9(x^2 - 6x) - 4(y^2 + 8y) = 19$$
$$9(x - 3)^2 - 4(y + 4)^2 = 19 + 81 - 64 = 36$$

Make the identification $X = x - 3$ and $Y = y + 4$ so the function can be written

$$9X^2 - 4Y^2 = 36 \quad \text{or} \quad \frac{X^2}{4} - \frac{Y^2}{9} = 1$$

Draw in the new axes with origin at $(3, -4)$. When $X = 0$, there are no real Y values. When $Y = 0$, $X = \pm 2$. Place these points on the graph. The asymptotes come out of the $y = \dots$ equation. Follow along the rearrangement to find the asymptote lines. (See Fig. 1-25.)

$$Y^2 = (9/4)X^2 - 9$$

$$Y = \pm \sqrt{(9/4)X^2 - 9}$$

For large values of X, $Y \approx \pm(3/2)X$. The addition of these asymptote lines allows completion of the graph.

Quick Tip

In graphing conics the first thing to determine is whether the equation is a circle, ellipse, or hyperbola. This is accomplished by looking at the numerical coefficients, their algebraic signs, and whether they are (numerically) different. Knowing the curve, the analytical techniques begin by looking for the values of x when y = 0, and the values of y when x = 0. The answers to these questions give the intercepts for the circle and ellipse, and the square root of a negative number for one determines that the curve is a hyperbola. The addition of linear terms moves the conics up and down and sideways and almost always requires a completing the square type of analysis, complete with axis shifting.

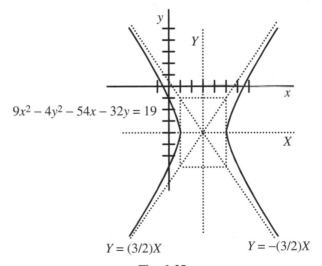

Fig. 1-25

If you can figure out what the curve looks like and can find the intercepts ($x = 0$ and $y = 0$) you are a long way toward graphing the function. The shifting of axes just takes attention to detail.

1-8 Graphing Trigonometric Functions

Graphing the trigonometric functions does not usually present any problems. There are a few pitfalls, but with the correct graphing technique these can be avoided. Before graphing the functions you need to know their general shape. The trigonometric relations are defined in an earlier section and their functions shown graphically. If you are not very familiar with the shape of the sine, cosine, and tangent functions draw them out on a 3 × 5 card and use this card as a bookmark in your text or study guide and review it every time you open your book (possibly even more often) until the word sine projects an image of a sine function in your mind, and likewise for cosine and tangent.

Let's look first at the sine function $y = \sin\theta$ and its graph in Fig. 1-26. The θ, called the argument of the function, is cyclic in 2π; whenever θ goes from 0 to 2π the sine function goes through one cycle. Also notice that there is a symmetry in the function. The shape of the curve from 0 to $\pi/2$ is mirrored in the shape from $\pi/2$ to π. Similarly, the shape of the curve from 0 to π is mirrored in the shape from π to 2π. In order to draw the complete sine curve

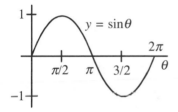

Fig. 1-26

we only need to know the points defining the first quarter cycle. This property of sine curves that allows construction of the entire curve if the points for the first quarter cycle are known will prove very valuable in graphing sine functions with complex arguments. Operationally, the values of the function are determined by "punching them up" on a hand calculator.

Example 1-32 Graph $y = 2\sin x$.

Solution: The 2 here is called the *amplitude* and simply scales the curve in the y direction. It is handled simply by labeling the y-axis, as shown in Fig. 1-27.

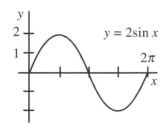

Fig. 1-27

Example 1-33　Graph $y = \cos 2x$.

Solution:　The phrase "cos" describes the general shape of the curve, the unique cosine shape. The $2x$ is the hard part. Look back at the basic shape of the cosine curve and note that when $\theta = \pi/2$, the cosine curve has gone through 1/4 of its cycle. The values of x for the points where $2x$ is zero and $\pi/2$ define the first quarter cycle. (One-quarter of a cycle is all that is necessary to graph the function.) To graph this function (y vs. x) we need to know only those values of x where the argument of the function ($2x$) is zero and $\pi/2$. The chart in Fig. 1-28 shows the values necessary for graphing the function.

$2x$	x	$\cos 2x$
0	0	1
$\pi/2$	$\pi/4$	0

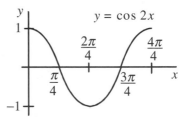

Fig. 1-28

Quick Tip

Do not start this chart with values of x; start with values for 2x. Read the previous sentence again. It is the key to correctly graphing trigonometric functions. Notice that the points on the x-axis are written as multiples of the first quarter cycle. It is a cumbersome way of writing the points, but it helps prevent mistakes in labeling the x-axis.

Go back over the logic of graphing trigonometric functions in this way. It is the key to always getting them graphed correctly. As the functions become more complicated, the utility and logic of this approach will become more evident.

$x/3$	x	$\sin (x/3)$
0	0	0
$\pi/2$	$3\pi/2$	1

Example 1-34　Graph $y = 2\sin(x/3)$.

Solution:　This is a sine function: the general shape of which can be seen clearly in your mind's eye. The amplitude of 2 is no problem. The argument $x/3$ requires setting up a chart to find the values of x defining the first quarter cycle of the sine function. Numbers associated with the argument of the function, the 3 (in the denominator) in this case, define the *frequency* of the function. While interesting in some contexts, knowing the frequency is not important in graphing. (Discussion of the frequency of periodic functions is contained in *Physics for the Utterly Confused*.)

Remember, in setting up the chart set $x/3$ equal to zero and solve for x. The sine of zero is zero. Next set $x/3$ equal to $\pi/2$ and solve for x. The sine of $\pi/2$ is 1. These two points define the first quarter cycle of the function. The remainder of the function is drawn in (Fig. 1-29) using the symmetry properties of sine functions.

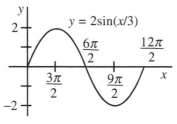

Fig. 1-29

Example 1-35   Graph $y = \sin(2x + \pi)$.

Solution: The introduction of the π in the argument of the function is the final complication in graphing trigonometric functions. This constant in the argument is called the *phase* and the sign of this constant moves the function to the left or right on the x-axis. It is not necessary to remember which sign moves the function which way. The placement of the function on the x-axis comes out of the analysis.

$2x + \pi$	x	$\sin(2x + \pi)$
0	$-(\pi/2)$	0
$\pi/2$	$-(\pi/4)$	1

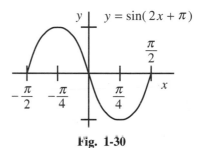

Fig. 1-30

Figure 1-30 shows a sine function with amplitude 1. The 2 affects the frequency and the moves the function right or left. Set up the π chart again forcing the argument to be zero or $\pi/2$ and determining the appropriate x value. Set $2x + \pi = 0$ and solve for $x = -(\pi/2)$. Set $2x + \pi = \pi/2$ and solve for $x = -(\pi/4)$. Draw the graph starting with the first quarter cycle of the sine function in the region from $-(\pi/2)$ to $-(\pi/4)$.

Example 1-36   Graph $y = (1/3)\cos(2x - \pi/3)$.

Solution: The function shown in Fig. 1-31 has another little twist to it, which has to do with the minus sign.

Follow the development of the chart and take $2x - \pi/3 = 0$ for the first point. This point is $x = \pi/6$ or $2\pi/12$.

$2x - \pi/3$	x	$\cos(2x - \pi/3)$
0	$\pi/6 = 2\pi/12$	1
$\pi/2$	$5\pi/12$	0

The next point is for $2x - \pi/3 = \pi/2$. This (second) point is then at $x = 5\pi/12$.

Set up the x-y coordinate system and place the first quarter of the cosine function between $2\pi/12$ and $5\pi/12$. With this section of the cosine function complete, draw in the remainder of the curve.

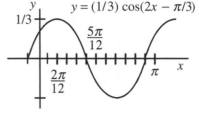

Fig. 1-31

Example 1-37   Graph $y = \tan(x - \pi/4)$.

Solution: If you are at all unfamiliar with the tangent function go back and review it in the trigonometry section. The important features as far as graphing is concerned are that $\tan\theta$ is zero when θ is zero and $\tan\theta$ is 1 when θ is $\pi/4$. The tangent curve goes infinite when θ goes to $\pi/2$, but a point at infinity is not an easy one to deal with.

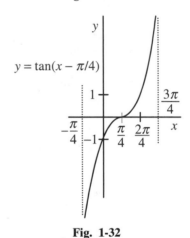

Fig. 1-32

For the function shown in Fig. 1-32, set up a chart and find the values of x that make $x - \pi/4$ equal zero and $\pi/4$. These two points allow construction of the function.

$x - \pi/4$	x	$\tan(x - \pi/4)$
0	$\pi/4$	0
$\pi/4$	$2\pi/4$	1

Solve $x - \pi/4 = 0$ for $x = \pi/4$.

Solve $x - \pi/4 = \pi/4$ for $x = \pi/2 = 2\pi/4$.

Be careful while graphing the tangent function, especially this one. This tangent function is zero when $x = \pi/4$, and 1 when $x = 2\pi/4$. The standard mistake is to take the function to infinity at $x = 2\pi/4$.

It's a Wrap

✔ Solve quadratic equations either by factoring or formula

✔ Keep the image of the sine, cosine, and tangent functions in your mind

✔ Graph trigonometric functions using the advanced techniques

✔ Switch from rectangular to polar coordinate systems and vice versa

✔ Know the basic laws for manipulating logarithms and exponents

✔ Evaluate and graph functions

✔ Graph circles (equal and positive coefficients of the x^2 and y^2 terms)

✔ Graph ellipses (nonequal and positive coefficients of the x^2 and y^2 terms)

✔ Graph hyperbolas (nonequal and one "−" sign of the x^2 and y^2 terms)

PROBLEMS

Test Yourself

1. Solve by factoring $x^2 - 3x - 10 = 0$.
2. Solve by factoring $2x^2 - 7x + 4 = 0$.
3. Solve by quadratic formula $x^2 + 7x - 3 = 0$.
4. Solve by quadratic formula $2x^2 + 5x - 6 = 0$.
5. Write $(a + b)^3$ using the binomial expansion.
6. Convert $\pi/4$ rad to degrees.
7. Convert 0.45 rad to degrees.
8. Convert 2.6 rad to degrees.
9. Convert 80° to radians.
10. Convert 200° to radians.
11. Switch the point $(-4,6)$ to polar form.
12. Switch the point $(2,-5)$ to polar form.
13. Switch 3 @ 20° to rectangular form.
14. Switch 5 @ 60° to rectangular form.
15. Solve $6 = \ln 3x$ for x.
16. Solve $e^x = 4.3$.
17. Solve $e^{2x} = 6.8$.
18. Write $\log 3 = 0.48$ in exponential form.
19. Solve $10^x = 0.56$.
20. Evaluate $\log(3.6)(4.1)^3$.
21. Evaluate $\log[(4.2)^2/2.3]$.
22. What is the y intercept for the function $y = 3x - 2$?
23. What is the y intercept for the function $y = x^2 + 3$?
24. Graph $y = x^2 - 2x - 8$.
25. Graph $y = x^2 + 6x + 9$.
26. Graph $(x - 2)^2 + (y + 3)^2 = 9$.
27. Graph $x^2 - 2x + y^2 + 4y = -1$.
28. Graph $4x^2 = -25y^2 + 100$.
29. Graph $9x^2 - 4y^2 = 36$.
30. Graph $4y^2 = 16x^2 + 64$.
31. Graph $y = 2 \sin(2x - \pi/2)$.
32. Graph $y = \tan(x + \pi/2)$.
33. Graph $y = \cos(3x - \pi/3)$.

ANSWERS

1. $(x - 5)(x + 2) \Rightarrow x = 5, x = -2$

2. $(2x - 1)(x - 3) \Rightarrow x = 1/2, x = 3$

3. $x = \dfrac{-7 \pm \sqrt{49 - 4(1)(-3)}}{2} = \dfrac{-7 \pm \sqrt{61}}{2}$

4. $x = \dfrac{-5 \pm \sqrt{25 - 4(2)(-6)}}{2(2)} = \dfrac{-5 \pm \sqrt{73}}{4}$

5. $(a + b)^3 = \dfrac{a^3}{0!} + \dfrac{3a^2b}{1!} + \dfrac{3 \cdot 2ab^2}{2!} + \dfrac{3 \cdot 2 \cdot 1b^3}{3!} = a^3 + 3a^2b + 3ab^2 + b^3$

6. $\dfrac{\pi}{4}$ rad $\dfrac{360°}{2\pi \, \text{rad}} = 45°$

7. $0.45 \, \text{rad} \dfrac{360°}{2\pi \, \text{rad}} = 25.8°$

8. $2.6 \, \text{rad} \dfrac{360°}{2\pi \, \text{rad}} = 149°$

9. $80° \dfrac{2\pi \, \text{rad}}{360°} = 1.40$ rad

10. $200° \dfrac{2\pi \, \text{rad}}{360°} = 3.49$ rad

11. This problem requires a diagram. There are two problems: first remember that the angle has to be measured counter-clockwise from the positive x-axis, and second be careful how you measure the angle. The angle measured from the $-x$ axis is determined by $\tan^{-1}(6/4) = 56.3°$. Now subtract $56.3°$ from $180°$ for $123.7°$. This is the correct number for the angle. The magnitude is the square root of 6 squared plus 4 squared or square root of $52 \approx 7.2$. Written in polar form the coordinates are $7.2 @ 123.7°$.

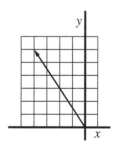

12. This problem requires a diagram. There are two problems: first remember that the angle has to be measured counterclockwise from the positive x-axis, and second be careful how you measure the angle. If you try and take the inverse tangent of 2 divided by -5, you will get a negative number—very confusing. The safest way to do this problem is to find the small angle between the y-axis and the arrow. Take the inverse tangent of 2 over 5 and get $22°$. Now add this $22°$ to $270°$ for $292°$.

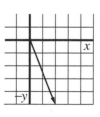

The magnitude of the arrow is the square root of 2 squared and 5 squared, or square root of $29 \approx 5.4$. Written in polar form the coordinates are 5.4 @ 292°.

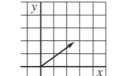

13. A diagram is helpful. Look at the diagram and write:
$x = 3\cos 20° = 2.8$
$y = 3\sin 20° = 1.0$

14. Picture the diagram in your mind and carefully calculate the x and y values.
$x = 5\cos(-60°) = 2.5$
$y = 5\sin(-60°) = -4.3$

15. $6 = \ln 3 + \ln x$, $6 - \ln 3 = \ln x$, $1.90 = \ln x$, $x = e^{1.9} = 6.7$

16. $\ln(4.3) = x$, $x = 1.46$

17. $\ln(6.8) = 2x$, $x = 0.96$

18. $10^{0.48} = 3$

19. $\log 0.56 = x$, $x = -0.25$

20. $\log(3.6) + 3\log(4.1) = 2.39$

21. $2\log(4.2) - \log(2.3) = 0.88$

22. -2

23. $+3$

24. This is a parabola that opens up. The function factors are: $y = (x - 4)(x + 2)$, so the curve crosses the x-axis at 4 and -2. The symmetry line is at $x = 1$.

25. This is a parabola that opens up and it is a perfect square: $y = (x + 3)^2$. The curve rests on the x-axis at the point $x = -3$ and is symmetric about the line $x = -3$.

26. This is a circle of radius 3, and it is displaced to the "new" coordinate lines of $x = 2$ and $y = -3$.

27. This is also a circle but it is a little more complicated. Complete the square. $(x - 1)^2 + (y + 2)^2 = -1 + 1 + 4 = 4$. This circle has radius 2 and is centered on the point $(1, -2)$.

28. This is an ellipse. Divide by 100, set first x, then y equal to zero and find the shape for the ellipse.

29. This is a hyperbola. First divide by 36. The curve goes through the points $(2,0)$ and $(-2,0)$ and does not exist at $y = 0$. The asymptote lines are from $y \approx (3/2)x$.

30. This is a hyperbola. Divide by 64. The curve goes through the points $(0,4)$ and $(0, -4)$ and does not exist at $x = 0$. The asymptote lines are from $y = \pm 2x$.

31. Set $2x - \pi/2 = 0$ and solve for x. This is the first point on the curve. Set $2x - \pi/2 = \pi/2$ and solve for x. This is the second point on the curve. Fill in the remainder of the sin curve from these two points.

32. Set $x + \pi/2 = 0$ and solve for x. This is the first point on the curve. Set $x + \pi/2 = \pi/4$ and solve for x. This is the "45°" point on the tangent curve. With these points fill in the tangent curve.

33. Set $3x - \pi/3 = 0$ and solve for x. This is the first point for the cos curve. Set $3x - \pi/3 = \pi/2$ and solve for x. This is the point needed to complete the first quarter of the cos curve. Then fill in the rest of the cos curve.

CHAPTER 2

LIMITS AND CONTINUITY

Do I Need to Read This Chapter? You should read this chapter if you need to review or you need to learn about

➡ The concept of a limit

➡ Algebraic techniques for finding limits

➡ Discontinuities

➡ Using discontinuities in graphing

The concept of the limit in calculus is very important. It describes what happens to a function as a particular value is approached. The derivative, one of the major themes of calculus, is defined in limit terms. This short chapter will help you to think in terms of limits. The first thing to understand about limits is that a limit of a function is not the value of the function. The change in thinking (from value to limit) is important because most functions are understood as a series of mathematical operations that can be evaluated at certain points simply by substitution.

The (polynomial) function $y = x^2 + 2x + 3$ can be evaluated for any real number: replace x with the number and perform the indicated operations. Asking the limit of this function as x approaches 2, for example, is an uninteresting question. The function can be evaluated at 2 or any point arbitrarily close to 2 by substituting and performing the operations.

Other functions, such as polynomial fractions, cannot be evaluated at certain points and these functions are best understood by thinking in terms of limits. The function $y = (x^2 - 4)/(x + 2)$ can be evaluated for any real number except -2. Replacing x by -2 produces the meaningless statement 0/0. Remember that any number times 0 is 0, but any number divided by 0 is "meaningless" (including 0/0). Looking at the limit of the function, as x approaches -2, tells us about the function in the vicinity of -2. The limit of the function is a convenient phrase for the question, "What happens to the function as a certain value is approached?" Writing this in mathematical notation we get the following:

$$\lim_{x \to -2} \frac{x^2 - 4}{x + 2} = \lim_{x \to -2} \frac{(x + 2)(x - 2)}{x + 2} = \lim_{x \to -2}(x - 2) = -4$$

The notation in front of the functions is read *the limit, as x approaches minus two*. In the case of rational functions, factoring and reducing the fraction helps in finding the limit.

Finding the limit of this function as $x \to -2$ helps in understanding the function. Since the original function gives the meaningless 0/0 at the point where $x = -2$, the function cannot exist, "does not have meaning," at $x = -2$. Graphing the function illustrates this point. The (simplified) function $y = x - 2$ is a straight line of slope 1 and intercept -2. The function $y = (x^2 - 4)/(x + 2)$ is also a straight line of slope 1 and intercept -2, but it does not exist at the point where $x = -2$. This nonexistence at $x = -2$ is illustrated on the graph in Fig. 2-1 with the open circle.

Example 2-1 Find the limit of $y = (x^2 + x - 2)/(x - 1)$ as $x \to 1$.

Solution: At $x = 1$ the fraction is $0/0$, so perform some algebra on the fraction before taking the limit.

$$\lim_{x \to 1} \frac{x^2 + x - 2}{x - 1} = \lim_{x \to 1} \frac{(x + 2)(x - 1)}{x - 1} = \lim_{x \to 1}(x + 2) = 3$$

As an exercise graph the original function, showing the nonexistence at $x = 1$.

Another category of function that is understood with the help of limits is polynomial fractions, where the higher power polynomial is in the denominator rather than the numerator. The simplest function to look at is $y = 1/x$. (The product of two variables equaling a constant describes certain relationships. For example, pressure and volume for a fixed amount of gas at constant temperature is described by $pV = $ const; the cost of comparable real estate times the commuting distance from a major commercial center is described by $RD = $ const.)

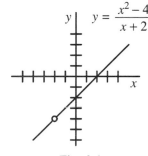

Fig. 2-1

This relationship $xy = 1$ or $y = 1/x$ is best understood in the context of its graph. Numbers can be assigned to x, and y values calculated but note how the concept and language of limits make graphing so much easier. Refer to Fig. 2-2 during this discussion. First consider positive values. The point $x = 1$, $y = 1$ is so easy to calculate it should not be ignored. The curve goes through this point. Now as x is made a larger and larger positive number, y approaches zero, but remains positive. This can be expressed in a simple sentence.

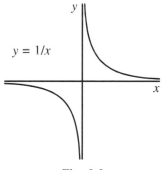

Fig. 2-2

As x approaches plus infinity, y approaches zero, but remains positive or in mathematical symbolism,

$$\text{As } x \to +\infty, y \to 0+$$

Write the situation for small values of x directly in mathematical symbolism,

$$\text{As } x \to 0+, y \to +\infty$$

What we are saying here is that if x is a very, very, very small number, even smaller than 0.000000001, the 1 divided by this number is a very, very, very large number. So as $x \to 0+$, $1/0+ \to +\infty$.

With this information, the positive portion of the graph can be drawn. In the case of pressure and volume or cost of real estate and distance, the problem dictates only positive values. In the function $y = 1/x$ no such restriction exists.

Refer to the graph in Fig. 2-2 and follow the logic and symbolism in the statements

$$\text{As } x \to 0-, \ y \to -\infty \quad \text{and} \quad \text{As } x \to -\infty, \ y \to 0-$$

Example 2-2   Graph the function $y = 1/(x - 1)$ using limit concepts and notation.

Solution: At $x = 1$, the function has value 1/0, which is hard to interpret. Using the limit concept the behavior of the function as x approaches 1 is easily understood. Note first that if x is greater than 1, the function is positive, and if x is less than 1, the function is negative. Apparently the function behaves differently as $x = 1$ is approached from either the positive or negative side. Remember that on the number line, positive is to the right and negative is to the left. In taking the limit it is necessary, in this case, to specify the direction of approach to 1. Notice how this is done in the notation.

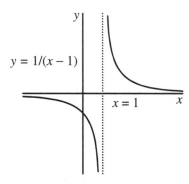

Fig. 2-3

$$\lim_{x \to 1(pos)} \frac{1}{x - 1} = \infty \qquad \lim_{x \to 1(neg)} \frac{1}{x - 1} = -\infty$$

Based on experience with $y = 1/x$, this function has the same shape; it is just displaced (or translated) 1 unit to the right. In $y = 1/x$, $x = 0$ is the asymptote line, but in $y = 1/(x - 1)$, $x = 1$ is the asymptote line. Follow the logic of the limit calculations and verify the graph as shown in Fig. 2-3.

Example 2-3   Graph $y = 1/x^2$ using limit concepts and notation.

Solution: Think limits and write the symbolic statements.

$$\text{As } x \to +\infty, \ y \to 0+ \ \text{As } x \to 0+, \ y \to +\infty$$

$$\text{As } x \to 0-, \ y \to +\infty \ \text{As } x \to -\infty, \ y \to 0+$$

Now draw in the curve. (See Fig. 2-4.)

Example 2-4 Discuss the function $y = \dfrac{x + 2}{x - 5}$ in the vicinity of $x = 5$.

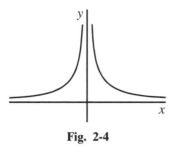

Fig. 2-4

Solution: The numerator of the function presents no problem. Even at $x = -2$, the function is $0/-7 = 0$, perfectly understandable. Based on past experience, the $x - 5$ in the denominator produces a vertical asymptote at $x = 5$. Place the asymptote line on the graph. Now, using limit language, describe the behavior of the function in the vicinity of $x = 5$.

$$\text{As } x \to 5+, \ y \to +\infty \quad \text{As } x \to 5-, \ y \to -\infty$$

There is an additional complication as x becomes large, either positive or negative. For large x the function becomes large number over large number. If, however, the fraction is multiplied by $1/x$ over $1/x$ the limit can be calculated easily:

$$\lim_{x \to \infty} \frac{x - 2}{x + 5}\left[\frac{1/x}{1/x}\right] = \lim_{x \to \infty} \frac{1 - 2/x}{1 + 5/x} = 1$$

This limit produces a horizontal asymptote. When x is greater than 5 (refer to the original function statement), the fraction is positive so this horizontal asymptote is approached from the positive side. When x is less than 5, but greater than -2, the function is negative. At $x = 0$, $y = -2/5$. For values of x less than (to the left of) -2, the function is always positive and for larger and larger negative x, the function approaches the limit 1 from the negative side. Go through the logic and verify the graph of Fig. 2-5.

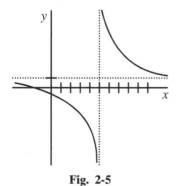

Fig. 2-5

As the powers of the polynomials increase, the functions become harder to graph. In Chapter 4 more complicated polynomials will be graphed with the aid of calculus.

Example 2-5 Find the limit of the function $y = \dfrac{3x^2 + 2x + 1}{x^2 + x + 1}$ as x goes to infinity.

Solution: Attempting to evaluate the function for large x produces the result large number over large number. Taking the limit with a little inventive algebra (multiplying the fraction by $1/x^2$ over $1/x^2$) produces

$$\lim_{x \to \infty} \frac{3x^2 + 2x + 1}{x^2 + x + 1}\left[\frac{1/x^2}{1/x^2}\right] = \lim_{x \to \infty} \frac{3 + 2/x + 1/x^2}{1 + 1/x + 1/x^2} = 3$$

Example 2-6 Find the limit of $\dfrac{x^4 + 3x^2}{x^5 + 2}$ as x goes to infinity.

Solution: Again use a little inventive algebra. Factor an x^4 out of the numerator and an x^5 out of the denominator:

$$\lim_{x \to \infty} \frac{x^4 + 3x^2}{x^5 + 2} = \lim_{x \to \infty} \frac{x^4(1 + 3/x^2)}{x^5(1 + 2/x^5)} = \lim_{x \to \infty}\left[\frac{1}{x}\right]\left[\frac{1 + 3/x^2}{1 + 2/x^5}\right] = 0$$

The first fraction has limit zero and the second limit 1. The product is zero.

This problem illustrates a manipulative rule for limits. $\lim_{x \to c} A \cdot B = [\lim_{x \to c} A][\lim_{x \to c} B]$

Similarly $\lim_{x \to c} \dfrac{A}{B} = \dfrac{\lim_{x \to c} A}{\lim_{x \to c} B}$ and $\lim_{x \to c} A^n = [\lim_{x \to c} A]^n$

Continuous functions are defined mathematically, usually over specific intervals. The requirements of a continuous function are: (1) it exists at every point in the defined interval and (2) the limit exists at every point and is equal to the value of the function at that point. Operationally, continuous functions are ones that can be drawn without lifting your pencil.

A discontinuous function is one that either: (1) doesn't exist at some point over the defined interval or (2) the limits from the positive and negative directions are different. Examples 2-1 through 2-4 are examples of discontinuous functions.

Another often used sample of a discontinuous function is the integer function, $y = [x]$, where the $[x]$ symbolism is read as *the largest integer contained in x.* For example, the largest integer contained in 2 is 2. The largest integer contained in 2.99 is 2. Add as many 9s as you like and the largest integer is still 2. The limit of the function as x approaches 3 from the negative side (slightly less than 3) is 2. The limit of the function as x approaches 3 from the positive side (slightly greater than 3) is 3. This discontinuity at each integer is shown in Fig. 2-6.

Look back over Examples 2-1 through 2-4 and note that the discontinuity occurs at the vertical asymptote.

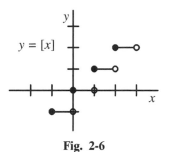

$y = [x]$

Another example of a discontinuous function is one defined on certain intervals such as

$$f(x) = \begin{cases} x + 3 \text{ for } -3 \leq x < 0 \\ 1.5x + 4 \text{ for } x \geq 0 \end{cases}$$

Fig. 2-6

This is a discontinuous function. Though it is defined everywhere over the interval, the limit as zero is approached from the positive side is 4, and the limit as zero is approached from the negative side is 3. The function is shown graphically in Fig. 2-7.

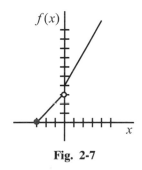

$f(x)$

An example of a function that is discontinuous from one side only is the square root function. The function $y = \sqrt{x}$ is not defined for negative x because there are no real square roots of negative numbers. In mathematical symbolism

Fig. 2-7

$$\lim_{x \to 0+} \sqrt{x} = 0 \quad \text{and} \quad \lim_{x \to 0-} \sqrt{x} \text{ does not exist}$$

The function is continuous to zero from the right side but not from the left side.

The cube root function, $y = \sqrt[3]{x}$, behaves differently. There are real cube roots of both positive and negative numbers as well as zero. This function is continuous over the entire range of real numbers for x.

It's a Wrap

✔ A limit is not the value of a function

✔ Limits are often different depending on the approach

✔ Discontinuities and limits can help in graphing

✔ Functions can be discontinuous

Test Yourself

PROBLEMS

1. Find $\lim_{x \to 1} \dfrac{x - 2}{x + 1}$.

2. Find $\lim_{x \to 0^+} \dfrac{1}{x^4}$.

3. Discuss the $\lim\limits_{x \to 2} \dfrac{x^2 - 4}{x - 2}$.

4. Find $\lim\limits_{x \to 1+} \sqrt{4x - 4}$.

5. For the previous problem discuss the limit as x approaches 1 from the negative side.

6. Discuss the $\lim\limits_{x \to 2} \dfrac{x^3 - 8}{x - 2}$.

7. Discuss the $\lim\limits_{x \to 1} \dfrac{x^3 - 1}{x^2 - 1}$.

8. Find $\lim\limits_{x \to \infty} \dfrac{6x^2 + 5x + 100}{3x^2 - 9}$.

9. Find $\lim\limits_{x \to \infty} \dfrac{10x^2 + 300x + 1}{5x + 1}$.

10. Use limit concepts to graph $y = \dfrac{1}{x - 3}$.

11. Discuss the function $y = \dfrac{x + 1}{x - 3}$ in the vicinity of $x = 3$.

12. Sketch the graph of $y = \dfrac{1}{x^4}$ using limit concepts.

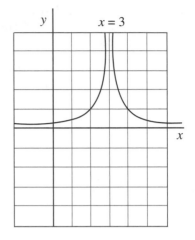

ANSWERS

1. $-\dfrac{1}{2}$

2. $+\infty$

3. The limit does not exist at $x = 2$. However, rewriting the fraction as $\lim\limits_{x \to 2}(x + 2)$, the limit is 4.

4. As x approaches 1 from the positive side the square root is of a smaller and smaller positive number. This square root is still positive, so the limit is zero.

5. If x approaches 1 from the negative side (less than 1) the square root is of a negative number and the limit does not exist. Neither are there any values of the square root for numbers less than 1.

6. The fraction does not exist for $x = 2$. It would be 0/0. But factoring or long dividing the fraction becomes $\lim\limits_{x \to 2}(x^2 + 2x + 4) = 12$.

7. This problem is similar to the one just above. The fraction does not exist for $x = 1$. Factor and note that a factor of $(x - 1)$ is in both the numerator and denominator making the fraction $\lim\limits_{x \to 1} \dfrac{(x^2 + x + 1)}{(x + 1)} = \dfrac{3}{2}$.

8. Divide up and down by $1/x^2$ and the fraction is 2 in the limit.

9. Divide up and down by $1/x$ and the denominator is a number while the numerator goes to infinity so the limit is infinity.

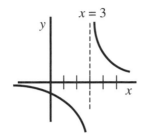

10. This is a graph like $y = 1/x$ but shifted so the vertical asymptote is at $x = 3$, where the function does not exist.

11. Again, this function has a discontinuity at $x = 3$. The graph is a little more complicated and requires a few points in the vicinity of $x = 3$.

   At $x = 2, y = -3$
   At $x = 1, y = -2$
   At $x = 4, y = 5$
   At $x = 5, y = 3$

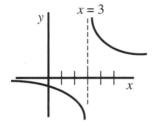

12. This graph is much like the one for $y = 1/x^2$. Since the powers are 2 and 4, the symmetry will be the same.

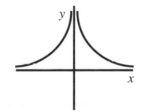

CHAPTER 3

DERIVATIVES

 You should read this chapter if you need to review or you need to learn about

➡ Definition of the derivative

➡ Derivatives of polynomials

➡ Product and quotient rules

➡ Trig, exponential, and Log functions

➡ Implicit differentiation and the chain rule

➡ Finding limits with derivatives

The derivative of a function is the slope of that function anywhere the function is well behaved. A function is well behaved in a region where there is a unique slope at every point. A constant function, $y = 2$, $y = -3$, is a straight line of slope zero. A linear equation, $y = 2x - 3$, $y = -x + 5$, has a constant slope (2 and -1 in these cases).

The simplest function that does not have a constant slope is the quadratic, $y = x^2$ (see Chapter 1 for a discussion of quadratics). The slope of the quadratic, considering only positive values of x, increases as x increases. Look at a magnified portion of the $y = x^2$ curve and approximate the slope of the curve at any point by writing a general expression for the slope of the straight line connecting two points x and $x + \Delta x$. The notation Δx means a small change in x so the point $x + \Delta x$ is very close to x.

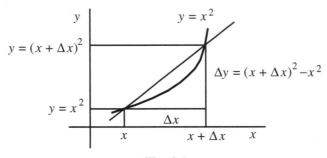

Fig. 3-1

Figure 3-1 shows the curve $y = x^2$ and the straight line connecting the points $(x, y = x^2)$ and $(x + \Delta x, y = (x + \Delta x)^2)$. The slope of the line between these adjacent points is

$$\frac{(x + \Delta x)^2 - x^2}{\Delta x}$$

The general expression for the slope of this curve at any point x is the limit of this approximate slope as Δx goes to zero. Using the mathematical symbolism of limits, the general expression for the slope of $y = x^2$ is

$$\frac{dy}{dx} = y' = f'(x) = \lim_{\Delta x \to 0} \frac{(x + \Delta x)^2 - x^2}{\Delta x}$$

This defining equation for the derivative is called dy/dx, where the d notation indicates the limit of $\Delta y/\Delta x$, or y', or $f'(x)$. For the quadratic, we have the following:

$$\frac{dy}{dx} = \lim_{\Delta x \to 0} \frac{(x^2 + 2x\Delta x + (\Delta x)^2) - x^2}{\Delta x} = \lim_{\Delta x \to 0} (2x + \Delta x) = 2x$$

This general expression for the derivative is used to determine the slope of the curve $y = x^2$ at any point. When $x = 3$, the function has value 9 and slope 6. When $x = 4$, the function has value 16 and slope 8. Another, more general, way of writing this definition is

$$\frac{dy}{dx} = \lim_{\Delta x \to 0} \frac{y(x + \Delta x) - y(x)}{\Delta x} \tag{3-1}$$

where the expression $y(x + \Delta x)$ means the value of y at $x + \Delta x$ and $y(x)$ means the value of y at x.

Example 3-1 Use the definition of the derivative to find the derivative of $y = x^3$.

Solution: Follow the definition of the derivative in Equation 3-1.

$$\frac{dy}{dx} = \lim_{\Delta x \to 0} \frac{(x + \Delta x)^3 - x^3}{\Delta x} = \frac{x^3 + 3x^2\Delta x + 3x\Delta x^2 + \Delta x^3 - x^3}{\Delta x}$$

$$\frac{dy}{dx} = \lim_{\Delta x \to 0} (3x^2 + 3x\Delta x + \Delta x^2) = 3x^2$$

3-1   Polynomials

There is a pattern to these derivatives as illustrated in the adjacent table. The pattern suggests a general rule for differentiating polynomials.

If $f(x) = cx^n$, then $f'(x) = cnx^{n-1}$.

$f(x)$	$f'(x)$
const	0
mx	m
x^2	$2x$
x^3	$3x^2$

Quick Tip

The general power law formula works for positive and negative exponents, and fractional exponents. For a function $f(x) = cx^n$, the derivative is $f'(x) = cnx^{n-1}$.

Example 3-2   Differentiate $y = 3x^3 + x + 2$.

Solution: Following the general definition of the derivative (Equation 3-1) write:

$$\frac{dy}{dx} = \lim_{\Delta x \to 0} \frac{[3(x + \Delta x)^3 + (x + \Delta x) + 2] - [3x^3 + x + 2]}{\Delta x}$$

Looking at the parentheses, the 2s add to zero and the xs add to zero. The $(x + \Delta x)^3$ term is in the previous problem as well so

$$\frac{dy}{dx} = \lim_{\Delta x \to 0} \frac{3x^3 + 9x^2 \Delta x + 9x \Delta x^2 + 3\Delta x^3 + \Delta x - 3x^3}{\Delta x}$$

$$= \lim_{\Delta x \to 0} 9x^2 + 9x \Delta x + 3\Delta x^2 + 1$$

$$\frac{dy}{dx} = 9x^2 + 1$$

Look at the general power law rule (If $f(x) = cx^n$, then $f'(x) = cnx^{n-1}$) and notice that if this law were applied to each of the terms, first the $3x^3$, then the x, this result would be achieved.

The previous problem is an example of a simple rule: The derivative of a collection of terms is the sum of the derivatives of the individual terms. In mathematical language

If $f(x) = u(x) + v(x)$ then $f'(x) = u'(x) + v'(x)$, and

if $f(x) = u(x) - v(x)$ then $f'(x) = u'(x) - v'(x)$.

Example 3-3   Using the general power law rule and the sum and difference rules (above) find the derivative of $y = 4x^3 + 3x^2 - 2x - 3$.

Solution: $y' = 3(4x^2) + 2(3x) - 2(1) - 0 = 12x^2 + 6x - 2$

Example 3-4   Find the slope of $y = 4x^3 + 3x^2 - 2x - 3$ at $x = 2$.

Solution: Using the expression for y' from the previous problem we can solve as follows:

$$y'(2) = 12(2)^2 + 6(2) - 2 = 48 + 12 - 2 = 58$$

Example 3-5 The cost in dollars to manufacture x number of a certain item is $CM = 120 + 0.02x^2$. This relation is valid for up to 70 items (maximum capacity of the facility) per month. Find the cost to manufacture the 10th, 40th, and 70th items. This is called the marginal cost.

Solution: The general expression for the cost per item is the derivative of the cost function, $d(CM)/dx$.

$$\frac{d(CM)}{dx} = 0.04x \qquad \frac{d(CM)}{dx}\bigg|_{10} = \$0.40 \qquad \frac{d(CM)}{dx}\bigg|_{40} = \$1.60 \qquad \frac{d(CM)}{dx}\bigg|_{70} = \$2.80$$

The derivative can be thought of as a rate. A most convenient way to illustrate this is with velocity and acceleration. One of the easiest rates to visualize is velocity, distance divided by time. If something moves 200 meters (m) in 50 seconds (s) we say it has a velocity of 4 m/s. This 200 m in 50 s produces an average velocity, $\Delta x / \Delta t$ in calculus language. The velocity at any instant during the 50 s may, however, be quite different from the average. To find the instantaneous velocity we first need to know how x varies with time, or $x = f(t)$. Then dx/dt, the limit as the time interval becomes shorter and shorter, is an expression for the instantaneous velocity, v, that can be evaluated at any time.

If something is changing velocity as it moves then we can take the difference in velocity between the beginning and end of a time interval and calculate the average acceleration over that time interval. The instantaneous acceleration $(a = dv/dt)$, the rate at which the velocity changes, is the derivative of the velocity-time relation evaluated at any time.

Example 3-6 Find the expression for the instantaneous velocity for the distance-time function $x = kt^3 - lt^2 + mt$ and evaluate the velocity at $t = 1$s. Take $k = 2$ m/s^3, $l = 4$ m/s^2, and $m = 5$ m/s.

Solution: The general expression for velocity is the time derivative of the $x = f(t)$, or

$$v = \frac{dx}{dt} = 3kt^2 - 2lt + m$$

and the velocity evaluated at $t = 1$ is

$$v|_{t=1} = (3)(2 \text{ m/s}^3)(1 \text{ s})^2 - (2)(4 \text{ m/s}^2)(1 \text{ s}) + 5 \text{ m/s} = 3 \text{ m/s}$$

Example 3-7 Continue Example 3-6 by finding the acceleration at $t = 2\,\text{s}$.

Solution: The general expression for the instantaneous acceleration is the time derivative of the expression for v. (See the previous problem for v.)

$$a = \frac{dv}{dt} = 2 \cdot 3kt - 2l$$

and the acceleration at $t = 2\,\text{s}$ is

$$a|_{t=2} = (6)\,(2\,\text{m/s}^3)(2\,\text{s}) - 2(4\,\text{m/s}^2) = 16\,\text{m/s}^2$$

Velocity and acceleration problems are excellent test problems. Be sure you know that given position as a function of time $x = f(t)$, the velocity is the first derivative, and the acceleration is the second derivative.

Given $x = 4 + 6t - 5t^2$ know how to find velocity ($v = 6 - 10t$) and acceleration ($a = -10$) and be able to evaluate velocity and acceleration at any time.

The velocity is the first derivative of position, $v = dx/dt$. The acceleration is the first derivative of the velocity, $a = dv/dt$. Both derivatives come from the same function. The velocity is the first derivative and the acceleration the second derivative. It is common to write v as a first derivative and a as a second derivative:

$$v = \frac{dx}{dt} \text{ and } a = \frac{d^2x}{dt^2} = \frac{d}{dt}\left(\frac{dx}{dt}\right)$$

You will encounter second and third derivatives of the same function in other areas.

Example 3-8 The gross domestic product (GDP) of a certain country is $N(t) = t^2 + 3t + 80$ in billions of dollars when t is measured in years. This growth was valid from 1980 to 1990. What was the rate of growth in 1986?

Solution: The growth rate is the derivative of the growth function.

$$\frac{dN}{dt} = 2t + 3$$

and evaluating for 1986,

$$\frac{dN}{dt}\bigg|_6 = 2(6) + 3 = 15 \text{ billion dollars per year.}$$

While the rate of growth is important the percentage of change is also important. In an absolute sense, a change of 50 in the manufacture of 500 items is a 10% change while a change of 50 in 50,000 items is a 0.1% change. Change is relative to the number of items manufactured. A 0.1% change will probably require no change in manufacturing but a 10% change might have a significant effect on manufacturing procedures. In order to get a measure of this relative change we make the following definition.

This *percentage rate of change* is defined as

$$\text{Percentage rate of change} = 100 \frac{\text{rate of change}}{\text{size of quantity}}$$

$$\text{Percentage rate of change} = 100 \frac{f'(x)}{f(x)}$$

Example 3-9   For the previous problem find the percentage rate of change in 1986.

Solution: The numerator of the fraction, dN/dt from the previous problem is:

$$\frac{dN}{dt}\bigg|_6 = 15 \text{ billions per year}$$

The actual number for 1968 is:

$$N(t) = t^2 + 3t + 80$$
$$N(6) = 6^2 + 3(6) + 80$$
$$N = 134 \text{ billion}$$

The percentage rate of change is then,

$$100 \frac{15}{134} = 11\% \text{ per year}$$

In 1986 the GDP is 134 billion, the rate of growth is 15 billion per year and the percentage rate of growth is 11% per year.

3-2   Product and Quotient Rule

Having established the derivatives of polynomials and worked some sample problems let's move on to fractions and products.

Example 3-10   Find the derivative of the function $y = x^2(x + 2)$.

Solution: $\dfrac{dy}{dx} = \lim\limits_{\Delta x \to 0} \dfrac{(x + \Delta x)^2(x + \Delta x + 2) - x^2(x + 2)}{\Delta x}$

$$\frac{dy}{dx} = \lim\limits_{\Delta x \to 0} \frac{(x^3 + x^2\Delta x + 2x^2 + 2x^2\Delta x + 2x\Delta x^2 + 4x\Delta x + x\Delta x^2 + \Delta x^3 + 2\Delta x^2) - (x^3 + 2x^2)}{\Delta x}$$

$$\frac{dy}{dx} = 3x^2 + 4x$$

Second Solution: The solution could have been obtained much easier by applying the rule for differentiating a product. The derivative of a product is the first term times the derivative of the second term plus the second term times the derivative of the first term. In mathematical symbolism,

$$\text{If} \quad f(x) = u(x)v(x) \quad \text{then} \quad f'(x) = u(x)v'(x) + u'(x)v(x)$$

For this problem, then,

$$\frac{dy}{dx} = x^2 \frac{d}{dx}(x + 2) + (x + 2)\frac{d}{dx}x^2 = x^2(1) + (x + 2)(2x) = 3x^2 + 4x$$

Quick Tip

In words the product rule is first $\times$ d (second) + second $\times$ d (first).

A similar, though somewhat more complicated rule applies for fractions. Again, use the basic definition of the derivative to find the differential of a fraction and see how the differentiation can be performed much easier with the fraction rule.

Example 3-11 Use the basic definition of the derivative to find dy/dx of $y = x + 1/x^2$.

Solution:

$$\frac{dy}{dx} = \lim_{\Delta x \to 0} \frac{\dfrac{x + \Delta x + 1}{(x + \Delta x)^2} - \dfrac{x + 1}{x^2}}{\Delta x}$$

$$\frac{dy}{dx} = \lim_{\Delta x \to 0} \frac{x^2(x + \Delta x + 1) - (x + 1)(x^2 + 2x\Delta x + \Delta x^2)}{(x + \Delta x)^2(x^2)(\Delta x)}$$

$$\frac{dy}{dx} = \lim_{\Delta x \to 0} \frac{x^3 + x^2\Delta x + x^2 - (x^3 + 2x^2\Delta x + x\Delta x^2 + x^2 + 2x\Delta x + \Delta x^2)}{(x + \Delta x)^2(x^2)(\Delta x)}$$

$$\frac{dy}{dx} = \frac{-x^2 - 2x}{x^4} = -\frac{x + 2}{x^3}$$

Second Solution: The general rule for differentiating a fraction is:

$$\text{If} \quad f(x) = \frac{u(x)}{v(x)} \quad \text{then} \quad f'(x) = \frac{v(x)u'(x) - u(x)v'(x)}{[v(x)]^2}$$

By applying this rule the problem becomes much easier:

$$\frac{dy}{dx} = \frac{x^2 \dfrac{d}{dx}(x + 1) - (x + 1)\dfrac{d}{dx}(x^2)}{(x^2)^2} = \frac{x^2(1) - (x + 1)(2x)}{x^4}$$

$$= \frac{-x^2 - 2x}{x^4} = -\frac{x + 2}{x^3}$$

Quick Tip

In words the quotient rule is top $\times d$ (bottom) $-$ bottom $\times d$ (top) all over (bottom)^2.

3-3   Trigonometric Functions

There is no gereral rule for determining the derivative of trigonometric functions. Each trigonometric function has a unique derivative. One will be done to demonstrate the approach. Consult a table of derivatives for the derivative of each specific trigonometric function.

Example 3-12   Apply the basic definition to find the derivative of the sine function, $y = \sin\theta$.

Solution: $\dfrac{dy}{d\theta} = \lim\limits_{\Delta\theta\to 0} \dfrac{\sin(\theta + \Delta\theta) - \sin\theta}{\Delta\theta}$

The $\sin(\theta + \Delta\theta)$ can be replaced with the sum of two angles identity (see earlier trigonometric identities).

$$\frac{dy}{d\theta} = \lim_{\Delta\theta\to 0} \frac{\sin\theta\cos\Delta\theta + \cos\theta\sin\Delta\theta - \sin\theta}{\Delta\theta}$$

As $\Delta\theta$ goes to zero, $\cos\Delta\theta$ goes to 1 (see the graph of the cosine function) so the problem reduces to

$$\frac{dy}{d\theta} = \cos\theta \lim_{\Delta\theta\to 0} \frac{\sin\Delta\theta}{\Delta\theta}$$

As $\Delta\theta$ approaches zero, $\sin\Delta\theta$ approaches $\Delta\theta$. For small angles the sine function looks like a straight line of slope 1. Check this out with your hand calculator.

Take the sine of 0.2 (rad), a little over 11°, and notice how close the sine of 0.2 is to 0.2. Now decrease the angle to 0.1, 0.01, 0.001, until your calculator no longer displays a difference between the sine and the angle. The limit of $\sin\Delta\theta$ over $\Delta\theta$ as $\Delta\theta$ goes to zero is 1, so the derivative of the sine function is the cosine function.

The approximation $\sin\theta = \theta$ for small θ is used in many problems in physics and engineering.

3-4   Implicit Differentiation

The general procedure for differentiating a polynomial function $y = x^2 + 2x$ is to apply the power law rule to each term and write $dy/dx = 2x + 2$. Another and often very convenient way of looking at the problem would be to differentiate the entire equation term by term, $dy = 2xdx + 2dx$, and then write $dy/dx = 2x + 2$. You should notice that most differential tables are written in this manner. As functions become more complicated differentiating term by term like this and implicit differentiation become more convenient. Suppose you have a function $x^4 + x^2y^2 - xy^3 = 18$ where it is impossible to solve for x in terms of y or y in terms of x. Implicit differentiation is the only way to find dy/dx.

Example 3-13 Find dy/dx for $x^4 + x^2y^2 - xy^3 = 18$ by implicit differentiation.

Solution: The x^2y^2 and xy^3 terms are treated as products.

$$4x^3dx + x^2d(y^2) + y^2d(x^2) - xd(y^3) - y^3dx = 0$$

$$4x^3dx + x^2(2ydy) + y^2(2xdx) - x(3y^2dy) - y^3dx = 0$$

Separate out the terms multiplying dy and dx.

$$(2x^2y - 3xy^2)dy = -(4x^3 + 2xy^2 - y^3)dx$$

and solve for dy/dx.

$$\frac{dy}{dx} = -\frac{4x^3 + 2xy^2 - y^3}{2x^2y - 3xy^2}$$

Example 3-14 Continue Example 3-13 by finding the value of the slope at $x = 2, y = 1$

Solution: The dy/dx is from the previous problem so

$$\left.\frac{dy}{dx}\right|_{\substack{x=2 \\ y=1}} = -\frac{4(2)^3 + 2(2)(1) - 1}{2(2)^2(1) - 3(2)(1)} = -\frac{32 + 4 - 1}{8 - 6} = -\frac{35}{2} = -17.5$$

The graph of $x^4 + x^2y^2 - xy^3 = 18$ goes through the point $(2, 1)$ and has a slope at this point of -17.5.

Example 3-15 Factory output in units of product output is $Q = 2x^3 + x^2y + y^3$ where x is the number of skilled workers and y is the number of unskilled workers. Current production output requires 30 hours of x's and 20 hours of y's. If one skilled worker is added how many unskilled workers can be removed while maintaining constant output?

Solution: Though it is not necessary, calculate the present output.

$$Q = 2(30)^3 + (30)^2(20) + (20)^3 = 80,000 \text{ units}$$

The problem requires an implicit differential. The Q is constant so that differential is zero and it was not necessary to calculate the present output.

$$6x^2 dx + 2xy dx + x^2 dy + 3y^2 dy = 0$$
$$(x^2 + 3y^2)dy = -(6x^2 + 2xy)dx$$
$$\frac{dy}{dx} = -\frac{6x^2 + 2xy}{x^2 + 3y^2}$$

Since we are asking the question about the numbers of workers the equation can be conveniently written in Δ format.

$$\Delta y = -\frac{6x^2 + 2xy}{x^2 + 3y^2}\Delta x$$

This statement now tells us how many unskilled workers can be removed if one skilled worker is added. And this is for constant output. Put in the numbers.

$$\Delta y = -\frac{6(30)^2 + 2(30)(20)}{(30)^2 + 3(20)^2}\Delta x = -3.1$$

Adding one skilled worker replaces three unskilled workers.

3-5   Change of Variable

Implicit differentiation and a change of variable become essential when functions become complicated and more than one rule is needed to perform a differentiation.

Example 3-16   Find the derivative of $y = (x + 3)^{25}$.

Solution: You could try to find someone to raise $x + 3$ to the 25th power or you could view the $x + 3$ as a variable and apply the power rule. Implicit differential also helps to simplify the problem.

$$dy = 25(x + 3)^{24} d(x + 3) \quad \text{so} \quad \frac{dy}{dx} = 25(x + 3)^{24}$$

Second Solution: Instead of just thinking of the $x + 3$ as the variable you can define a new variable, $u = (x + 3)$ so the function reads $y = u^{25}$ with implicit derivative $dy = 25u^{24} du$. The derivative of u is from the definition of u, so $du = dx$ and

$$dy = 25(x + 3)^{24} dx \quad \text{and} \quad \frac{dy}{dx} = 25(x + 3)^{24}$$

Example 3-17 Find the derivative of $y = \cos^3(x^2 + 2)$.

Solution: From any Mathematical Table, the derivative of the cosine is negative sine. Use the power rule first to obtain $dy = 3\cos^2(x^2 + 2)d\cos(x^2 + 2)$. View the $x^2 + 2$ as the variable and take (the derivative of)

$$d\cos(x^2 + 2) = -\sin(x^2 + 2)d(x^2 + 2) = -2x\sin(x^2 + 2)dx$$

Combining,

$$\frac{dy}{dx} = [3\cos^2(x^2 + 2)][-2x\sin(x^2 + 2)] = -6x\cos^2(x^2 + 2)\sin(x^2 + 2)$$

Second Solution: Notice how much easier and less susceptible to error the problem becomes when a change of variable is made early on in the problem. First set $u = x^2 + 2$ so $du = 2xdx$. Now write the problem as $y = \cos^3 u$ and differentiate implicitly.

$$dy = 3(\cos^2 u)d\cos u = 3(\cos^2 u)(-\sin u)du$$

$$= 3[\cos^2(x^2 + 2)][-\sin(x^2 + 2)]2xdx$$

$$\frac{dy}{dx} = -6x\cos^2(x^2 + 2)\sin(x^2 + 2)$$

Example 3-18 Use a change of variable approach to find the derivative of $y = (x^2 + 3x + 1)^3$

Solution: Let $u = x^2 + 3x + 1$ so the function reads $y = u^3$ with derivative $dy = 3u^2du$.

Find $du = 2xdx + 3dx$ and substitute so $dy = 3(x^2 + 3x + 1)^2(2x + 3)dx$ and the derivative is

$$\frac{dy}{dx} = 3(2x + 3)(x^2 + 3x + 1)$$

3-6   Chain Rule

In many practical situations a quantity is given in terms of a variable and then this variable is expressed in terms of a third variable. A problem may be described this way because the first variable is not easily written in terms of the third or perhaps it is conceptually easier to understand the process in two steps.

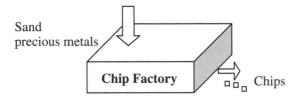

Suppose the cost of manufacturing a certain item, say a computer chip, depends on the number of items produced. The number of items produced depends on the length of time the "fab" facility operates to produce the chips, the length of time for the production run. If the cost per unit (dollars per chip) is dC/dN and the rate of production (chips per hour) is dN/dt, then cost per unit of time is the product of these two derivatives.

$$\frac{dC}{dt} = \frac{dC}{dN}\frac{dN}{dt}$$

Example 3-19 Find du/dt for $u = x^2 + 2x$ and $x = t^3 - 3$.

Solution: This requires a chain derivative: $\dfrac{du}{dt} = \dfrac{du}{dx}\dfrac{dx}{dt}$

$$\frac{du}{dt} = \frac{du}{dx}\frac{dx}{dt} = (2x + 2)(3t^2)$$

Example 3-20 The monthly profit for selling tacos follows:

$$P = -1200 + 30q - 0.002q^2$$

where q is the number of tacos sold. The relation between number of workers and tacos is $q = 100n$.

Find dP/dn, the profit per worker for $q = 1000$ tacos sold.

Solution: This problem requires a chain derivative: $\dfrac{dP}{dn} = \dfrac{dP}{dq}\dfrac{dq}{dn}$.

$$\frac{dP}{dq} = 30 - 0.004q \quad \text{and} \quad \frac{dq}{dn} = 100$$

dP/dn can now be written.

$$\frac{dP}{dn} = (30 - 0.004q)(100) = 3000 - 0.4q$$

Substitute for $q = 1000$ to determine the profit per worker at 1000 tacos sold per month.

$$\frac{dP}{dn}_{q=1000} = 3000 - 0.4(1000) = 2600$$

3-7   Logarithms and Exponents

The differentials of several logarithms and exponents are listed below.

If $y = e^x$ then $dy = e^x dx$

If $y = a^x$ and $a > 0$ and $a \neq 1$ then $dy - (\ln a)a^x dx$

If $y = \ln x$ then $dy = \frac{1}{x} dx$

If $y = \log_a x$ and $a > 0$ and $a \neq 1$ then $dy = \frac{1}{(\ln a)x} dx$

Example 3-21   Find the derivative of $y = e^{x^2-3}$.

Solution: Think, or write, $y = e^u$ and the derivative is

$$dy = e^u du = e^{x^2-3}(2x dx) \text{ or } \frac{dy}{dx} = 2xe^{x^2-3}$$

Example 3-22   Find dy/dx of $y = \ln x^2$.

Solution: $dy = \frac{1}{x^2} 2x dx$ or $\frac{dy}{dx} = \frac{2}{x}$

Example 3-23   Find the derivative of $y = x^{1.5} \ln(x^2 + 2)$.

Solution: This looks bad. But, if you proceed slowly, applying the rules one at a time, the differentiation is not all that difficult. The hard part is proceeding logically. This is a product so write

$$dy = x^{1.5} d[\ln(x^2 + 2)] + \ln(x^2 + 2)dx^{1.5}$$

The differential of $\ln(x^2 + 2)$ is, according to the table,

$$\frac{1}{x^2 + 2}d(x^2 + 2) = \frac{2xdx}{x^2 + 2}.$$

The differential of $x^{1.5}$ is $(1.5)x^{0.5}dx$.

Putting it all together we get the following:

$$dy = x^{1.5}\frac{2xdx}{x^2 + 2} + (1.5)x^{0.5}\ln(x^2 + 2)dx \quad \text{or} \quad \frac{dy}{dx} = \frac{2x^{2.5}}{x^2 + 2} + (1.5)x^{0.5}\ln(x^2 + 2)$$

Example 3-24   Find the derivative of $y = \tan x/x$.

Solution: Your first reaction to this problem probably is to apply a fraction rule. Apply the fraction rule.

$$\frac{dy}{dx} = \frac{x\frac{d}{dx}(\tan x) - (\tan x)\frac{d}{dx}x}{x^2} = \frac{x\sec^2 x - (\tan x)}{x^2}$$

Second Solution: Often viewing a fraction as a product makes for an easier differential. Switching to an implicit differential and viewing the problem as a product,

$$dy = x^{-1}d(\tan x) + (\tan x)d(x^{-1}) = x^{-1}\sec^2 x\,dx + (\tan x)(-x^{-2})dx$$

$$\frac{dy}{dx} = \frac{x\sec^2 x - \tan x}{x^2}$$

Example 3-25   Find the derivative of $y = e^{-x}\sin x$.

Solution: This is a product. Proceed methodically and the problem is not difficult.

$$dy = e^{-x}d(\sin x) + \sin x\,d(e^{-x}) = e^{-x}(\cos x)dx - e^{-x}(\sin x)dx$$

$$\frac{dy}{dx} = e^{-x}(\cos x - \sin x)$$

3-8 L'Hopital's Rule

There is another technique for finding the limit of an indeterminate form for a fraction where the limit looks like 0/0 or ∞/∞. The technique is known as L'Hopital's rule.

The rule is as follows:

For a fraction where the $\lim\limits_{x\to\infty}\dfrac{f(x)}{g(x)}$ is $\dfrac{0}{0}$ or $\dfrac{\infty}{\infty}$ the limit is determined by taking the derivative of the numerator and the derivative of the denominator separately and again taking the limit.

$$\text{If } \lim_{x\to\infty}\frac{f(x)}{g(x)} \text{ is indeterminate, take } \lim_{x\to\infty}\frac{f'(x)}{g'(x)}.$$

The rule can be applied multiple times if necessary. Be sure that each time the rule is applied both the numerator and the denominator are differerentiated.

Example 3-26   Find $\lim\limits_{x\to\infty}\dfrac{x}{x+1}$.

Solution: This is the indeterminate form ∞/∞ so apply L'Hopital's rule.

$$\lim_{x\to\infty}\frac{1}{1}=1$$

Second Solution: This answer can also be seen by applying algebra to the problem: multiplying up and down by $1/x$.

$$\lim_{x\to\infty}\frac{x}{x+1}=\lim_{x\to\infty}\frac{x(1/x)}{(x+1)(1/x)}=\lim_{x\to\infty}\frac{1}{1+1/x}=1$$

Example 3-27   Find $\lim\limits_{x\to\infty}\dfrac{x-3}{x^2+5}$.

Solution: Apply L'Hopital's rule: $\lim\limits_{x\to\infty}\dfrac{1}{2x}=0$

Example 3-28   Find $\lim\limits_{x\to1}\dfrac{x^5-3x^3+5x-3}{4x^5+2x^3-5x^2-1}$.

Solution: Substituting $x=1$ into this fraction produces the indeterminate form $0/0$.

So apply L'Hopital's rule: $\lim\limits_{x\to1}\dfrac{5x^4-9x^2+5}{20x^4+6x^2-10x}=\dfrac{4}{16}=\dfrac{1}{4}$.

It's a Wrap

✔ For polynomials follow $dy = cnx^{n-1}dx$

✔ For products: first d (second) + second d (first)

✔ For quotients: bot d (top) – top d (bot) over bot squared

✔ For two descriptive equations use the chain rule

✔ If the variables cannot be separated use implicit derivative

✔ Use tables for trig, exponential, and log functions

✔ Apply L'Hopital's rule to indeterminate forms

Test Yourself

PROBLEMS

1. Find the derivative of $y = 4x^3 + 3x^{-1}$.
2. Find the derivative of $y = (2x^3 + 1)(x^2 - 2)$.
3. Find the derivative of $y = \dfrac{x^2 + x + 1}{(x + 2)}$.
4. Find the derivative of $y = (x^2 + 2x + 1)^3$.
5. The number of pet strollers sold in the United States in the 1995 to 2003 time period was: $n(t) = 0.04t^2 + 0.3t + 15$, where $n(t)$ is in thousands of carriages when n is the number of the year. The year 1995 corresponds to $t = 5$ and so on up to t = 13 for 2003. What was the rate of pet stroller sales in the year 2000?
6. If $y = u^3 + 3u^2 + 1$ and $u = x^2 + 2$ find dy/dx.
7. If $y = \dfrac{1}{u}$ and $u = 2x^2 + 1$ find dy/dx.
8. Find dy/dx for $xy^2 + 2x^2 + y^3 = 3$.
9. Find dy/dx for $(xy)^2 - 2y - 5x = 0$.
10. Find $f'(x)$ for $f(x) = \ln(x^2 + 2)$.
11. Take the derivative of $y = e^{x/3}$.
12. Differentiate $f(u) = \cos^2 u$.
13. Use L'Hopital's rule to find $\lim\limits_{x \to \infty} \dfrac{2x + 3}{4x}$.
14. Find $\lim\limits_{x \to 2} \dfrac{2x - 4}{x^2 + 3x - 10}$.

ANSWERS

1. $y' = 12x^2 - 3x^{-2}$

2. This is a product rule problem.
   $y' = (2x^2 + 1)d(x^3 - 2) + (x^3 - 2)d(2x^2 + 1)$
   $y' = (2x^2 + 1)(3x^2) + (x^3 - 2)(4x)$
   $y' = 10x^4 + 3x^2 - 8x$

3. This is a quotient rule problem.
   $$y' = \frac{(x + 2)d(x^2 + x + 1) - (x^2 + x + 1)d(x + 2)}{(x + 2)^2}$$
   $$y' = \frac{x^2 + 4x + 1}{(x + 2)^2}$$

4. This problem is best viewed as a change of variable. Write $y = u^3$.
   $y' = 3u^2du = 3(x^2 + 2x + 1)^2(2x + 2) = 3(2x^3 + 6x^2 + 6x + 2)$

5. The word rate in the problem is the key word. It tells you to differentiate.
   $dn/dt = 0.08t + 0.3$ and evaluating this derivative at $t = 10$ (year 2000)
   $dn/dt\big|_{10} = 0.8 + 0.3 = 1.1$ or 1100 pet strollers per year.

6. This is a chain derivative problem.
   $$\frac{dy}{dx} = \frac{dy}{du}\frac{du}{dx} = (3u^2 + 6u)(2x)$$
   $$= [3(x^2 + 2)^2 + 6(x^2 + 2)][2x] = 3x^4 + 18x^2 + 24$$

7. $\dfrac{dy}{dx} = \dfrac{dy}{du}\dfrac{du}{dx} = (-u^{-2})(4x) = -\dfrac{4x}{(2x^2 + 1)^2}$

8. $y^2dx + 2xydy + 4xdx + 3y^2dy = 0$
   $(2xy + 3y^2)dy = -(y^2 + 4x)dx$
   $$\frac{dy}{dx} = -\frac{y^2 + 4x}{2xy + 3y^2}$$

9. $2x^2ydy + 2xy^2dx - 2dy - 5dx = 0$
   $(2x^2y - 2)dy = (5 - 2xy^2)dx$
   $$\frac{dy}{dx} = \frac{5 - 2xy^2}{2x^2y - 2}$$

10. $f'(x) = \dfrac{dy}{dx} = \dfrac{1}{x^2 + 2}d(x^2 + 2) = \dfrac{2x}{x^2 + 2}$

11. $\dfrac{dy}{dx} = e^{x/3}d(x/3) = \dfrac{1}{3}e^{x/3}$

12. $f'(u) = 2(\cos u)d(\cos u) = -2(\sin u)(\cos u)$

13. $\lim\limits_{x \to \infty} \dfrac{2}{4} = \dfrac{1}{2}$

14. This is an indeterminate form for $x = 2$, so use L'Hopital's rule.

$$\lim\limits_{x \to 2} \frac{2}{2x + 3} = \frac{2}{7}$$

CHAPTER 4

GRAPHING

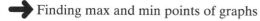

 You should read this chapter if you need to review or you need to learn about

➡ Graphing cubic and higher power curves

➡ Finding max and min points of graphs

➡ Identifying points of inflection

➡ Interpreting graphs of real problems

The next three chapters, graphing, max-min problems, and related rate problems, all deal with applications of the derivative. They are considered the most difficult topics in the first semester of calculus, particularly graphing.

Before going any farther in this chapter go back and review the graphing of parabolas, paying particular attention to visualizing the curve before plotting points and sketching the curve. Also go back and look over the concept of asymptotes in the chapter on limits. Many authors approach graphing in the calculus by using calculus only. We do not use that approach. Graphing is difficult enough without using exclusively new techniques. We use the graphing techniques of algebra; particularly those techniques discussed in the graphing of parabolas and higher power curves. Let's look at a couple of simple problems and see how the derivative can be used in curve sketching.

Example 4-1   Sketch the graph of $y = 4$.

Solution: This is a straight line parallel to the x-axis as shown in Fig. 4-1. Further, it is a horizontal line. The derivative of $y = 4$ is zero.

Any curve in the form $y = $ const. is a horizontal line parallel to the x-axis and has zero slope.

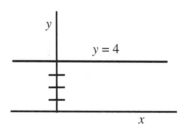

Fig. 4-1

Example 4-2   Sketch the graph of $y = 2x$.

Solution: The derivative of $y = 2x$ is 2. The slope is everywhere constant and equal to 2 (see Fig. 4-2). Any linear function has a constant derivative and a constant slope.

Example 4-3   Sketch the graph of $y = -x^2 - 2x + 8$.

Solution: This is a parabola (the 2 in the exponent) that opens down (the minus sign

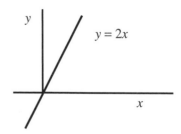

Fig. 4-2

in front of the squared term) and it is shifted up or down and sideways (the 8 means it is shifted up and down and the presence of an x term means it is shifted sideways). If you did not know this, go back and review the sections relating to graphing parabolas.

Factoring, $y = (-x + 2)(x + 4)$ tells us that the curve crosses the x-axis when $x = 2$ and $x = -4$. These are the values of x that make $y = 0$. Place these two points on the graph and with the knowledge that the curve opens down, expect a positive value of y at the symmetry line, $x = -1$. Substituting $x = -1$ into the original function produces $y = 9$. These points and the knowledge that the curve is a parabola are sufficient for drawing the sketch shown in Fig. 4-3.

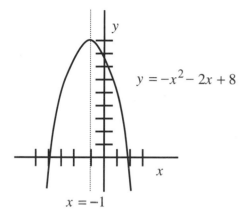

$$y = -x^2 - 2x + 8$$

$$x = -1$$

Fig. 4-3

There is another point that is so easy it is not worth passing up. Look at the original function and note that at $x = 0$, $y = 8$.

So, where does calculus come in? At the point $(-1, 9)$ the slope of the curve is zero. This means that the derivative must be zero at the point $x = -1$. When the derivative of a function is zero, the slope is zero and the curve is flat (at that point). Setting the derivative of $y = -x^2 - 2x + 8$ equal to zero should produce the value of $x = -1$.

The function $y = -x^2 - 2x + 8$ has derivative $y' = -2x - 2 = -2(x + 1)$. Setting $-2(x + 1) = 0$ produces the solution $x = -1$ and we already know $y = 9$ for $x = -1$.

How does calculus help in graphing? When the derivative of a parabola is zero, the curve has a ∩ or ∪ shape. Zero slope means the curve is flat and the only place where a parabola is flat is at a peak or a valley. Because there may be more than one peak or valley for a specific curve these points are called *relative maximum* or *relative minimum* points. The broader application of this approach is very helpful in higher (than 2) power curves such as the one in the next problem.

Example 4-4 Sketch the graph of $y = x^3 - 3x^2 + 2$.

Solution: The dominant term is x^3 so for large enough values of x the curve looks like a cubic. It goes up to the right and down to the left. If you have any trouble understanding that last sentence go back to the chapter on graphing and look up cubics. For $x = 0$, $y = 2$. With this most rudimentary analysis we know that the curve goes up to the right, down to the left, and passes through (0,2).

The derivative of a cubic is a quadratic, and a quadratic has two solutions or, in this case, two points where the slope is zero.

The derivative of $y = x^3 - 3x^2 + 2$ is $y' = 3x^2 - 6x = 3x(x - 2)$ with solutions $x = 0$ and $x = 2$. Substituting these values into the original function produces the points $(2,-2)$ and $(0,2)$. Place these points on the coordinate system, remembering that they are points on the curve where the slope is zero, and the curve is easily sketched.

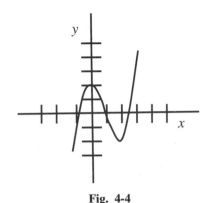

Fig. 4-4

The point (1,0) is easy to calculate. And if more detail is desired the values of (2,2) and $(-1,-2)$ can be obtained easily. These last two points show an approximate position where the curve crosses the x-axis (Fig. 4-4).

Quick Tip

A third power curve has a second power derivative. The second power derivative has at most two points, solutions, where the derivative is zero. A fourth power curve has a third power derivative and at most three points where the derivative is zero and so on for higher power curves. The number of points where a polynomial has zero slopes is at most equal to one less than the power of the polynomial. There is, however, another little twist to this rule as illustrated in the following problem.

Example 4-5   Sketch the graph of $y = 3x^4 - 4x^3 + 1$.

Solution: This is a fourth degree equation. The $3x^4$ term dominates for large x so the curve eventually rises to the right and the left. The point (0,1) is easy. The point (1,0) is almost as easy. Now apply some calculus analysis. Differentiate the function, set the derivative equal to zero, and find where the curve has zero slope.

Differentiating the function, $y' = 12x^3 - 12x^2 = 12x^2(x - 1)$, and setting the derivative equal to zero, $12x^2(x - 1) = 0$, produces two values of x where the slope is zero, $x = 0$ and $x = 1$. We already have the coordinates of these points, and now we know the curve has zero slope at these points.

This analysis has produced a dilemma. How can the curve go up to the right, go up to the left, and have two points such as ∪ or ∩? It can't! One of the points where the slope is zero must be a point where the curve, going up or down, becomes flat and continues on up or down. The point $x = 1$ is lower than the point $x = 0$ so the point at $x = 1$ must be the one with shape ∪ and the point at $x = 0$ must be the one where the curve flattens out (Fig. 4-5). The exact shape in the vicinity of both $x = 0$ and $x = 1$ can be checked by trying some points in the original equation. There is, however, a better way. It involves calculus and it is easier.

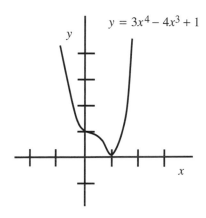

$y = 3x^4 - 4x^3 + 1$

Fig. 4-5

The first derivative of a function set equal to zero determines where the function has zero slope. At these points the curve is either concave up or concave down, or has an inflection point where the slope is zero. The second derivative of the function produces the answers here. To get a feel for how the second derivative works look at the previous problems.

Example 4-3 is the sketch of $y = -x^2 - 2x + 8$ and algebra analysis indicates a parabola that looks like ∩, symmetric about the line $x = -1$. The first derivative of y is $y' = -2x - 2 = -2(x + 1)$ and setting $y' = 0$ produces the point (for zero slope) of $x = -1$. The second derivative of $y = -x^2 - 2x + 8$ is $y'' = 2$. The second derivative is negative at $x = -1$ and in fact everywhere on the curve.

A simple parabola $y = x^2 \pm ()x \pm ()$ opens up (∪). The first derivative is $y' = 2x \pm ()$ and the second derivative is $y'' = 2$. For a parabola that opens up, the second derivative is positive at the minimum value.

Look at Example 4-4, the graph of $y = x^3 - 3x^2 + 2$. The first derivative $y' = 3x^2 - 6x$ produces zero slopes at $x = 0$ and $x = 2$. The second derivative $y'' = 6x - 6$ is negative at $x = 0$, and positive at $x = 2$.

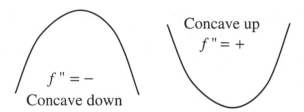

Both of these problems illustrate the rule that at points where the first derivative goes to zero, the curves are concave up when the second derivative is positive and concave down when the second derivative is negative. This is a calculus method of determining where the curves are concave up and where they are concave down.

Now let's take a look at Example 4-5, the one with the horizontal inflection point. The original function is $y = 3x^4 - 4x^3 + 1$ with first derivative $y' = 12x^3 - 12x^2$ producing $x = 0$ and $x = 1$ as the points where the slope is zero. The second derivative is $y'' = 36x^2 - 24x = 12x(3x - 2)$. At $x = 1$, the second derivative is positive indicating the curve is concave up at this point. At $x = 0$, the second derivative has value 0 indicating neither concave up nor concave down, but a point of inflection.

These three problems illustrate the use of calculus in graphing. What we have learned so far can be summarized as follows:

- Take the first derivative. Set this first derivative equal to zero and solve the resulting equation to find points where the curve has zero slope.

- Take the second derivative and evaluate the second derivative at the points where the slope is zero.

  If the second derivative is positive, the curve is concave up.

  If the second derivative is negative, the curve is concave down.

  If the second derivative is zero, the curve has a point of inflection.

Example 4-6   Sketch the graph of $y = x^3 + x^2 - 2x$.

Solution: The dominant term is x^3 so the curve eventually goes up to the right and down to the left. The point $x = 0$, $y = 0$ is easy. Before differentiating, note that the curve has zero slope at no more than two points because the highest power is 3. Follow along the rules as they are written above.

The first derivative is $y' = 3x^2 + 2x - 2$. Setting $3x^2 + 2x - 2 = 0$ results in a quadratic that cannot be factored so apply the quadratic formula

$$x = \frac{-2 \pm \sqrt{2^2 - 4(3)(-2)}}{2(3)} = \frac{-2 \pm \sqrt{28}}{6} = 0.55, -1.2$$

Figure 4-6 is a first cut at the graph. It is based only on knowing that the curve goes up to the right, down to the left, passes through (0,0), and has zero slope at $x = 0.55$ and $x = -1.2$.

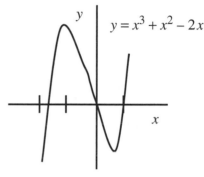

Fig. 4-6

Is it possible to easily find the points where the curve $y = x^3 + x^2 - 2$ crosses the x-axis? Maybe, maybe not, but it is at least worth trying a couple of obvious points:

At $x = 1$, $y(1) = 1 + 1 - 2(1) = 0$.

At $x = -1$, $y(-1) = -1 + 1 + 2 = 2$.

At $x = -2$, $y(-2) = -8 + 4 + 4 = 0$.

There is no point in trying further numbers. A cubic only crosses the x-axis at most three times and we have the three places where it crosses. Finding the y values at the turning points, where the slope is zero, may or may not be important to you. With this added information, the curve can be sketched as in Fig. 4-7.

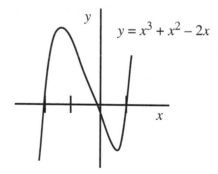

Fig. 4-7

Second Solution: There is another feature of this curve that can be analyzed using calculus. Look at the left part of the curve that looks like a parabola opening down and then the right part of the curve that looks like a parabola opening up. On this left part of the curve the slope becomes more and more negative until some point, between $x = -1.2$ and, $x = 0.55$, the slope of the curve, though still negative, starts becoming more positive. The point where this happens is also called a *point of inflection*. The strict definition of this point (of inflection) is that it is the point where the slope changes from becoming more negative to becoming more positive or vice versa.

Quick Tip

The analysis of points of inflection can be confusing, which is why these subtleties have been put off until now. There are two kinds of points of inflection, one where the curve goes to zero slope but does not have a ∪ or ∩ shape, and the other where the curve changes from having an increasingly negative slope to an increasingly positive slope.

The confusion does not end here, however. The first type of point of inflection is determined by evaluating the second derivative at the point where the first derivative goes to zero. The second type of point of inflection, called a *critical point*, is found by setting the second derivative equal to zero. Read this paragraph again and again until the distinction is clear in your mind.

The first derivative of $y = x^3 + x^2 - 2x$ is $y' = 3x^2 + 2x - 2$ and this resulted in the points $x = 0.55$ and $x = -1.2$, where the curve crossed the x-axis. The second derivative is $y'' = 6x + 2$, which is positive at $x = 0.55$ and negative at $x = -1.2$ confirming the previous analysis of this curve. The new feature is obtained by setting the second derivative equal to zero $2(3x + 1) = 0$. The second derivative is zero at $x = -1/3$. Look again at the curve in Fig. 4-7 and see that this is a very reasonable point for the curve to change slope from becoming more and more negative to becoming more and more positive. This is another graphing tool involving calculus.

Example 4-7   Sketch the graph of $y = 8x^5 - 5x^4 - 20x^3$.

Solution: This is a fifth degree curve so it increases (rises) rapidly with large positive x and goes rapidly negative for large negative values of x. The function factors to $y = x^3(8x^2 - 5x - 20)$ producing the point $x = 0$, $y = 0$.

The first derivative is

$$y' = 40x^4 - 20x^3 - 60x^2$$

$$y' = 20x^2(2x^2 - x - 3)$$

$$y' = 20x^2(2x - 3)(x + 1)$$

Setting the first derivative equal to zero ($y' = 0$) produces three points $x = 0$, $3/2$, and -1. These are the points where the curve has zero slope.

The second derivative is

$$y'' = 160x^3 - 60x^2 - 120x$$

$$y'' = 20x(8x^2 - 3x - 6)$$

To determine the shape of the curve where the slope equals zero, find y'' at each point: $y''(0) = 0$, horizontal point of inflection; $y''(3/2) = 225$, $\cup$ shape; $y''(-1) = -100$, $\cap$ shape.

The value of the function at each turning point is found by putting the values of x in the function.

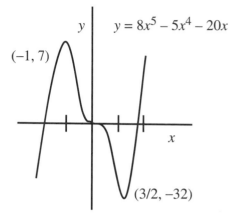

$y = 8x^5 - 5x^4 - 20x$

(-1, 7)

(3/2, -32)

Fig. 4-8

At $x = 0$, $y(0) = 0$

At $x = 3/2$, $y(3/2) = (3/2)^3[8(3/2)^2 - 5(3/2) - 20] \approx -32$

At $x = -1$, $y(-1) = -1(8 + 5 - 20) = 7$

All the inflection points, both horizontal and vertical, are found by setting $y'' = 0$:

$20x(8x^2 - 3x - 6) = 0$. The inflection points are at $x = 0$ and the solutions to $8x^2 - 3x - 6 = 0$ are

$$x = \frac{3 \pm \sqrt{9 - 4(8)(-6)}}{2(8)} = \frac{3 \pm 14.2}{16} = -0.70, 1.1$$

These are most reasonable points, being where we expect the points of inflection to occur. The function is sketched in Fig. 4-8.

So far we have looked at polynomials. This is the type of function you will encounter most often. Your course may or may not include the graphing of rational functions (polynomial fractions). Polynomial fractions introduce one more interesting twist to the use of derivatives in curve sketching, and that is: "What happens to a curve when the derivatives are undefined." This is best illustrated by example.

Example 4-8 Sketch the graph of $f(x) = \dfrac{x}{(x + 1)^2}$.

Solution: The denominator of $f(x)$ is zero at $x = -1$ meaning that the function does not exist there, so draw a vertical line at $x = -1$. For large x, the function looks like $1/x$. Look back to Chapter 2 and refresh your memory on what a $1/x$ curve looks like and keep this shape in mind throughout the rest of the analysis. An easy point to plot is (0,0). Now take the first derivative of the function.

$$f'(x) = \frac{(x + 1)^2(1) - x[2(x + 1)](1)}{(x + 1)^4}$$

$$f'(x) = \frac{(x + 1)(1 - x)}{(x + 1)^4}$$

$$f'(x) = \frac{1 - x}{(x + 1)^3}$$

The derivative is zero when $x = 1$. And when $x = 1$, $f(x) = 1/4$. The curve is flat at this point.

At this point in the analysis some "algebraic" analysis is in order. When x is less than -1, the function is negative and as x gets more and more negative the function looks like the $1/x$ curve. When x is between 0 and -1 the function is negative, and close to -1 the function is large negative.

The first derivative has shown that the curve is flat at (0,1/4) but the basic shape of the $1/x$ curve indicates that $f(x)$ looks like the $1/x$ curve for large x.

The curve to the right of the $x = -1$ line starts out very negative, passes through the point (0,0) is flat at (1,1/4) and then assumes the classic $1/x$ shape for large x. Now look at the slope situation. To the right of $x = -1$ the curve rises with negative slope passing through (0,0) goes flat at (1,1/4) and somewhere beyond $x = 1$ switches from a slope going more negative to a slope going more positive. This is a point of inflection that should come out of analysis using the second derivative.

Take the second derivative.

$$f''(x) = \frac{(x + 1)^3(-1) - (1 - x)[3(x + 1)^2(1)]}{(x + 1)^6}$$

$$f''(x) = \frac{(x + 1)^2[-(x + 1) - 3(1 - x)]}{(x + 1)^6}$$

$$f''(x) = \frac{2x - 4}{(x + 1)^4}$$

The second derivative is equal to zero when $x = 2$. $f(2) = 2/9$. The second derivative confirms the inflection in the slope and gives the point (2,2/9) as the specific point. With this information the curve can be sketched as shown in Fig. 4-9.

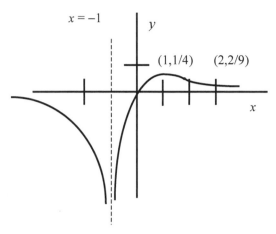

Fig. 4-9

Example 4-9 Sketch the graph of $y = x^2/(x - 1)$.

Solution: At $x = 1$ this function is undefined (1/0). Therefore, draw a dashed vertical line on the coordinate axes at $x = 1$ indicating that the curve may exist to the right or left of this line, but not on the line. There is no dominant term in the same manner as for polynomials but, applying similar reasoning, look what happens when x is a large positive or negative number. When x is large the $x - 1$ in the denominator looks like x and the function looks like $y \approx x$. In the language of the chapter on limits: As $x \to \pm\infty$, $y \approx x$.

Add a dashed line, $y = x$, to the coordinate axes remembering that this is an asymptote line.

Now apply some calculus analysis. The first derivative of the function is, using the quotient rule:

$$y' = \frac{(x - 1)(2x) - x^2}{(x - 1)^2} = \frac{2x^2 - 2x - x^2}{(x - 1)^2} = \frac{x^2 - 2x}{(x - 1)^2}$$

Before setting $y' = 0$, note that the derivative does not exist at $x = 1$. But we already knew that because the function does not exist at $x = 1$ so it is not surprising that the derivative does not exist there. Note, however, that as x approaches 1 from either the positive or negative side, the slope of the curve is negative. This information may be helpful in sketching the graph. (See Fig. 4-10.)

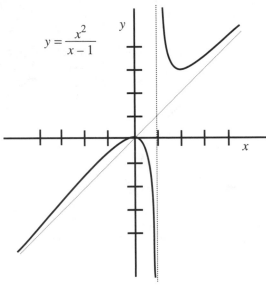

$$y = \frac{x^2}{x-1}$$

Fig. 4-10

Setting $y' = 0$ produces $x(x - 2) = 0$, and the two points where the slope equals zero are $x = 0$ and $x = 2$. The values of the function for these two points are: $y(0) = 0$ and $y(2) = 2^2/(2 - 1) = 4$ so the slope of the curve is zero at $(0,0)$ and $(2,4)$.

If you are unsure of the shape of the curve in certain regions, check a point. With the information generated from the calculus and the concepts of limits you should get very close to the correct curve. As you gain more confidence you will not resort to checking specific areas of the curve by testing a point.

The previous problem is typical of the more difficult ones you will encounter in your course. It is probably beyond what you will encounter on a test because of the complexity of the analysis and the potential for confusion. Sketches of the graphs of polynomials are much more popular as test problems. Know how to graph polynomials and you will be well along toward a good test score in graphing.

Having gone through examples of what you can expect to encounter in graphing problems, it is now time to write down some procedural guidelines for graphing curves of the general form $y = f(x)$.

Guidelines for Graphing with Calculus

1. Look for the dominant term. If the function is a polynomial, the highest-power term gives the shape of the curve for large positive or negative numbers

and one less than this highest power gives the maximum number of points where the curve has zero slope.

2. If the function is a fraction ask how it behaves for large x. Does it look like a straight line, a parabola, or what? Also look for places where the curve does not exist.

3. Take the first derivative. Set the first derivative equal to zero and solve for values of x where the curve has zero slope. Determine the y value at these points and add these points to the coordinate axes.

4. Take the second derivative. Evaluate the second derivative at the points where the slope is zero: If the second derivative is positive, the curve is concave up; if the second derivative is negative, the curve is concave down; if the second derivative is zero, the curve has a point of horizontal inflection.

5. Set the second derivative equal to zero and solve for values of x where the curve changes concavity. These are points where the slope of the curve changes from going more positive to going less positive or from going more negative to going less negative.

6. Sketch the curve. If you are unsure of the curve in certain places, plot a few points.

These are guidelines for graphing functions. You may not always need all of the steps listed here. Depending on what you are looking for in the problem, you may not need to perform each step in detail. These guidelines will, however, allow you to graph just about any function you encounter. Now it is time for some application problems.

Example 4-10 The sales for a certain consumer item are growing in a quadratic way with time while the discard rate remains a constant over time. Analysts expect this trend to continue for 5 years. The number of these items in the hands of consumers as a function of time is $N(t) = 3.2t^2 - 3t + 24$. The $3.2t^2$ term represents the quadratic growth in sales, the $-3t$ term represents the discard rate, and the 24 represents the number now in consumer hands. Sketch the graph of N vs. t. Determine if there is anything else in the graph or the calculus analysis that will help in business planning.

Solution: This function is a quadratic that opens up. It starts at $N = 24$ when $t = 0$. Only positive values of t have meaning. The equation cannot be factored so let's continue with the analysis remembering that we can always come back to the solution for $N = 0$ if necessary.

The first derivative of the function $N(t) = 3.2t^2 - 3t + 24$ is $N' = 6.4t - 3$.

Setting the first derivative equal to zero produces $6.4t - 3 = 0$ and the value $t = 0.47$ for zero slope of N vs. t. The value of the function at $t = 0.47$ is

$$N(0.47) = 3.2(0.47)^2 - 3(0.47) + 24 = 23.3.$$

The second derivative is 6.4, a positive number so the shape of the curve at $x = 0.47$, and everywhere is $\cup$. With this information the curve can be plotted as in Fig. 4-11. The curve never crosses the t-axis.

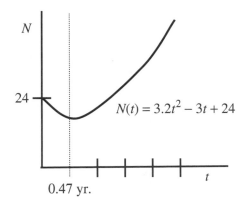

$N(t) = 3.2t^2 - 3t + 24$

0.47 yr.

Fig. 4-11

In addition to showing graphically the number of units predicted as needed, the first derivative tells us something else. For the first half year the number of items in consumer hands will decline (the minimum in the curve is at 0.47 year), then will rise. If the model is correct, suppliers need to be prepared for modest decrease followed by a much greater increase in demand for the product.

Example 4-11   The volume of lumber available in a managed forest follows the formula $V = (0.08)t^2 - (0.001)t^3$ over the first 60 years' life of the forest. Find the general shape of the curve from 0 to 60 years and determine the optimum time for harvesting the forest.

Solution: Only positive time from 0 to 60 years is interesting. The volume of lumber is in arbitrary units depending on the size of the forest. The curve starts out as a quadratic and then begins to flatten out with the growth of the t^3 term. This is reasonable. Trees grow rapidly in their early years and then slow down as they reach maturity.

Take the first derivative of the function $V = (0.08)t^2 - (0.001)t^3$ to obtain $V' = (0.16)t - (0.003)t^2$. Set this first derivative equal to zero to find the times when the curve has zero slope. $t(0.16 - 0.003t) = 0$ produces values of $t = 0$, and $t = 533$. The time of 533 years is well beyond where the formula is valid. The time $t = 0$ is very reasonable. The curve is flat at $t = 0$ and rises throughout the 60 years when the formula is valid.

Take a second derivative: $V'' = 0.16 - (0.006)t$ and set this equal to zero; $0.16 - (0.006)t = 0$ produces a value of $t = 0.16/0.006 = 27$ years. This second derivative test shows a change in concavity at 27 years. This means that the change in volume with time, the slope of the V vs. t curve, reaches a maximum at 27 years and after this time begins to drop off.

The most appropriate time to harvest this forest is at 27 years. A year or two more or less from this number won't make much difference because the slope is not changing rapidly around 27 years (Fig. 4-12).

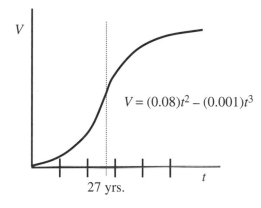

$$V = (0.08)t^2 - (0.001)t^3$$

27 yrs.

Fig. 4-12

Example 4-12   A certain disease is infecting an animal population. Experience with this disease shows that after injection with the appropriate antidote the number of animals infected with the disease follows the following formula:

$$P(t) = \frac{(20t + 8)}{(t^2 + 1)}$$

where t is measured in weeks. Find the time after the injection when the most animals will be affected by the disease and the total number affected. $P(t)$ is measured in thousands.

Solution: The t^2 term in the denominator ensures that as time goes on the number of infected animals will eventually tend to zero. If it did not, we should be looking for another antidote! There are no positive values of t where the curve does not exist. If this model correctly predicts the total number of animals affected by the disease and the time when this maximum occurs, then the antidote is working as predicted and we are assured that all the animals will eventually be cured shortly after the disease peaks.

At $t = 0$, $P(t) = 8$. This is when the antidote is administered to the animals. Finding the general shape of the P vs. t curve is ideally suited to calculus analysis.

The first derivative of $P(t) = \dfrac{20t + 8}{t^2 + 1}$ is

$$P' = \frac{(t^2 + 1)20 - (20t + 8)(2t)}{(t^2 + 1)^2} = \frac{-20t^2 - 16t + 20}{(t^2 + 1)^2}$$

Set the first derivative equal to zero and obtain

$$\frac{-5t^2 - 4t + 5}{(t^2 + 1)^2} = 0 \quad \text{or} \quad 5t^2 + 4t - 5 = 0$$

This equation cannot be factored, so solve by quadratic formula:

$$t = \frac{-4 \pm \sqrt{16 - 4(5)(-5)}}{2(5)} \frac{-4 \pm \sqrt{116}}{10}$$

$$t = 0.68, -1.5$$

Only the positive value has meaning so take $t = 0.68$ weeks for the zero slope condition and calculate P

$$P(0.68) = \frac{20(0.68) + 8}{(0.68)^2 + 1} = \frac{13.6 + 8}{1.46} = \frac{21.6}{1.46} = 14.8$$

The disease should peak at 0.68 week or 5 days after administration of the antidote with a maximum of 14.8 thousand animals infected on that day. After the fifth day, the number infected should decline as illustrated in Fig. 4-13.

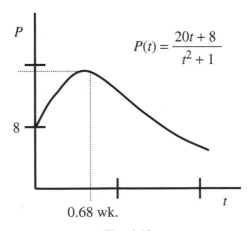

Fig. 4-13

Apply "algebraic" techniques along with calculus

 Use first derivative to find local max or min or inflections

✔ Use second derivative to determine concavity

✔ Use second derivative to find inflections in slopes

Test Yourself

PROBLEMS

1. Graph $y = x^2 + 2x - 8$.
2. Graph $y = x^3 - 3x^2 + 2$.
3. Graph $f(x) = x^4 + 8x^3 + 18x^2 - 8$.
4. Graph $y = x^3 - 3x^2 + 3x + 1$.
5. Graph $y = u + \dfrac{1}{u}$ for positive u.
6. Graph $y = \dfrac{x^2}{x - 2}$.

ANSWERS

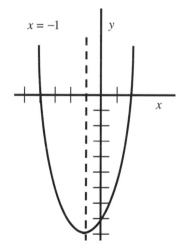

1. This is a parabola that opens up. The function factors to $y = (x + 4)(x - 2)$ so the curve crosses the x-axis at 2 and -4. $y(0) = -8$. The first derivative is $y' = 2x + 2 = 2(x + 1)$ with $y' = 0$ at $x = -1$. $y(-1) = (3)(-3) = -9$. This is the information necessary to draw the parabola.

2. This is a cubic with the cubic shape for large x. The curve goes through $(0,2)$. The first derivative is $y' = 3x^2 - 6x = 3x(x - 2)$ so the slope is zero at $x = 0$ and $x = 2$. $y(2) = 8 - 12 + 2 = -2$. $y'' = 6x - 6$ At $x = 0$, y'' is negative (maxima), and at $x = 2$, y'' is positive (minima). This is sufficient information to draw the curve.

3. This is a fourth degree equation so for large x it looks like a quadratic, but even more dramatic. $f(0) = -8$. The first derivative is $f'(x) = 4x^3 + 24x^2 + 36x = 4x(x + 3)^2$. The slope is zero when $x = 0$ and $x = -3$. $f(-3) = 19$. Take the second derivative: $f''(x) = 12x^2 + 48x + 36 = 12(x + 3)(x + 1)$. At $x = 0$, the second derivative is positive indicating a minimum. At $x = -3$, the

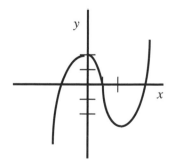

second derivative is zero indicating a point of inflection. The second derivative is zero when $x = -1$ and -3. The point at $x = -3$ is a horizontal inflection. (The first derivative is zero.) And the point at $x = -1$ must be where the curve changes from becoming more negative to more positive. (The first derivative is not zero at this point.) This is sufficient information to sketch the curve.

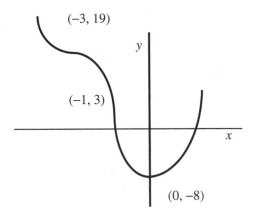

4. This is a cubic with the cubic shape for large x. At $x = 0$, $y = 1$. The first derivative is $y'' = 3x^2 - 6x + 3 = 3(x - 1)^2$. Slope is zero at $x = 1$; $y(1) = 2$. The second derivative is $y'' = 6x - 6 = 6(x - 1)$ and is equal to zero at $x = 1$, indicating an inflection point.

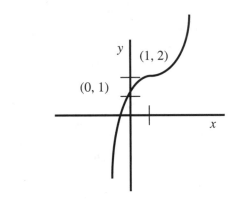

5. This curve is a combination of the classic $1/u$ shape combined with a linear term. For large u the curve approaches a straight line, and for small u the function is large. The line $y = u$ is an asymptote line. The curve can be sketched with this information but application of calculus provides other important features. The first derivative is $y' = 1 - 1/u^2$, and $y' = 0$ when $u^2 = 1$. The curve is flat at $u = 1$ and $y = 2$.

6. There is a vertical asymptote at $x = 2$, $y(0) = 0$. For large positive x the curve looks like $y \approx x$ and for large negative x the curve looks like $y \approx -x$.

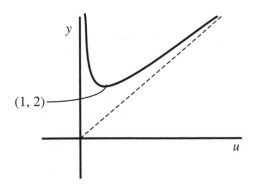

The first derivative is $y' = \dfrac{(x-2)(2x) - x^2(1)}{(x-2)^2} = \dfrac{x(x-4)}{(x-2)^2}$.

The slope is zero when $x = 0$ and 4. $y(4) = 8$.

Taking a second derivative looks complicated so look to an "algebraic" technique.

As $x \to 2(pos)$, $y \to +\infty$. As $x \to 2(neg)$, $y \to -\infty$.

This completes the analysis necessary to sketch the curve.

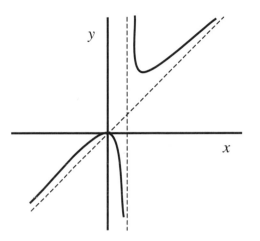

CHAPTER 5

MAX-MIN PROBLEMS

 Do I Need to Read This Chapter? You should read this chapter if you need to review or you need to learn about

➡ Designing containers for minimum volume

➡ Finding maximum areas for given constraints

➡ Designing beams for maximum strength

➡ Maximum production for agricultural crops

➡ Minimizing production costs

➡ Maximizing profit by price selection

Max-min problems are unique to calculus. As the name implies, a variable is maximized or minimized in terms of another variable. A typical problem would ask the question:

Max

"What is the maximum volume of a cylindrical container that can be made from a given amount of material?" The volume of the container is the variable to be maximized while the surface area of the container is limited by the amount of material allowed. In this example an equation for the volume $(V = \dots)$ is the defining equation. It defines the variable to be maximized, the volume, in terms of the dimensions of the container. The specification of a certain amount of material for the container is called the *constraint equation*. It relates the variables in the defining equation so the defining equation can be written in terms of one variable. This all becomes much clearer after a couple of problems.

Min

Once the defining equation is written in terms of one variable it is differentiated to find where the slope is zero. Where the slope of this curve is zero, the curve is at a maximum or a minimum. The value of the second derivative tells whether that point is a maximum or a minimum. Finding the points where the slope is zero and then identifying those points as either maximum ∩, or minimum ∪, has already been done in the graphing chapter. Max-min problems use much the same analysis techniques as with graphing.

Writing the defining equation is usually relatively easy. The hard part of max-min problems is finding the constraint equation and then doing the algebra so as to get the defining equation written in terms of one, other than the one to be maximized or minimized, variable and in as simple a form as possible.

There are very few max-min problems where the defining equation is written directly in terms of one variable. They are seen rarely on tests. They are considered too easy! Let's slowly go through a couple of max-min problems before setting down guidelines for working the problems and going on to the more challenging problems. Learn the procedure and max-min problems are not difficult.

Example 5-1   Design an open-top box for maximum volume. The box is to be made from a square piece of material of dimension a. (See Fig. 5-1.) What size square should be cut from each corner to make the box?

Solution: The side of the square taken from each corner is x. After the corner pieces are removed, the box is formed by bending the sides along the lines indicated.

The defining equation is $V = (a - 2x)^2 x$.

The bottom of the box is $a - 2x$ by $a - 2x$ and the height is x. The a is a constant, making the equation for V one with only one variable, x. Multiplying, we have the following:

$$V = (a^2 - 4ax + 4x^2)x = 4x^3 - 4ax^2 + a^2x$$

Differentiate V, and set the derivative equal to zero to find the maxima and minima of the curve of V vs. x.

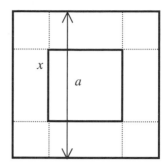

Fig. 5-1

$$V' = 12x^2 - 8ax + a^2 = a^2 - 8ax + 12x^2 = (a - 6x)(a - 2x)$$

Setting $V' = 0$ produces values for x of $a/6$ and $a/2$. These are the maxima or minima. The value $a/2$ is obviously the minimum since this is a box of zero volume! The value $a/6$ must be the maximum. The second derivative test will tell for sure. The second derivative of V is $V'' = 24x - 8a$. At $x = a/2$, $V'' = 12a - 8a = 4a$ (positive or minima), and at $x = a/6$, $V'' = 4a - 8a = -4a$ (negative or maxima).

Maximum volume occurs when the square piece removed from the edge of the original square is one-sixth the length of the side.

Example 5-2 A rectangular area is to be enclosed with 320 ft of fence (Fig. 5-2). What dimensions of rectangle give the maximum area?

Solution: The quantity to be maximized is the area, the product of the lengths of the two sides of the rectangle. The defining equation, the A equals . . . equation, is $A = ab$. Before maximizing the area (taking the derivative of A), the product ab must be written in terms of one variable. This requires a "constraint" equation relating a to b. The constraint in the problem is that the total length of fence $2a + 2b$ must be equal to 320. With this constraint equation A can be written in terms of a or b, it makes no difference.

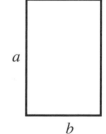

Fig. 5-2

Solve the constraint equation for a, and substitute in the area equation.

$$2a + 2b = 320 \quad \text{so} \quad a = 160 - b \quad \text{and} \quad A = (160 - b)b = 160b - b^2$$

The maximum occurs when the graph of A vs. b goes through a maximum. A maximum is defined, in calculus, as slope zero and second derivative negative.

Differentiating to find $\dfrac{dA}{db} = 160 - 2b$ and setting this equal to zero we get $160 - 2b = 0$ and $b = 80$.

The second derivative is $\dfrac{d^2A}{db^2} = -2$ confirming $b = 80$ as a maximum.

Go back to the constraint equation and note that for $b = 80$, $a = 80$. The area is maximum for a square.

Often max-min problems can be done with the first and second derivative. If you feel a little insecure, sketch the graph of the function. All the information, and then some, is already available for sketching the graph.

The original equation $A = 160b - b^2$ is a parabola that opens down and goes through the points $b = 0$ and $b = 160$ with symmetry line at $b = 80$. If you had any trouble with that last sentence, go back to the graphing of parabolas and review the procedure. The calculus tells us that the slope is zero at $b = 80$ and that the curve goes through a maximum at that point. This confirms what we already know from algebra analysis. The curve is sketched in Fig. 5-3.

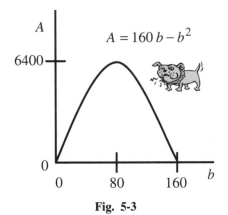

Fig. 5-3

This problem is an excellent pattern for max-min problems. Go through this problem again concentrating on the procedure, not the mathematics, and follow along the guidelines for doing max-min problems.

Guidelines for Max-Min Problems

1. Draw a diagram to help visualize the problem.
2. Write down the defining equation.
3. Tie the two variables in the defining equation together with a constraint equation.
4. Write the defining equation in terms of one variable.

5. Take the first and second derivatives to find maxima and minima.
6. Go back to the constraint equation and find all the quantities desired in the problem.

Example 5-3   Revenue as a function of price is given by the formula $R(p) = 200p - 8p^2$. Find the price for maximum revenue.

Solution: This revenue statement is realistic. For small price, p, the revenue rises linearly but as p increases sales fall off as indicated by the p squared term. Revenue should peak reaching a maximum when the R vs. p curve goes flat. So, take the first derivative and find the price for maximum revenue.

$R'(p) = 200 - 16p$ and setting this equal to zero $p = \dfrac{200}{16} = 12.50$. The second derivative is -16 confirming that the point at 12.50 is a maximum.

Maximum revenue occurs when the price is set at $12.50.

Example 5-4   The strength of a rectangular wooden beam varies jointly as the width and cube of the depth of the beam. Find the dimensions of the strongest beam that can be cut from a log of radius R.

Solution: Sketch the round log and the rectangular beam. Do you remember word problems in algebra that contained phrases like "... varies jointly as ... ?" This problem is included to remind you that some instructors use this language in calculus problems. The first statement in the problem, translated into algebra, is $S = wd^3$. This is the defining equation.

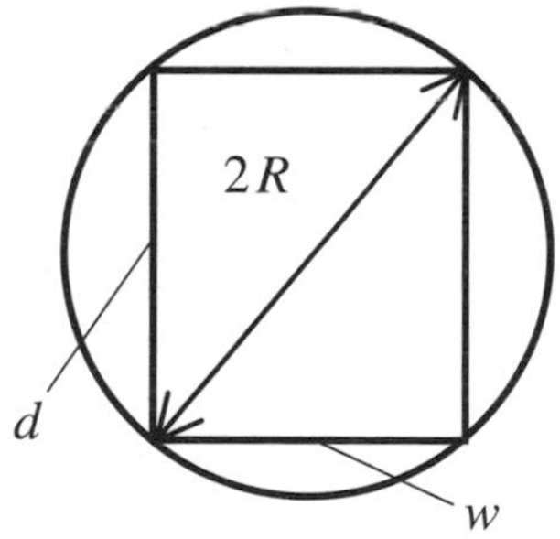

Fig. 5-4

To translate the problem statement completely, there should be a constant in front of the w but we are not going to calculate specific strengths, just the dimensions for maximum strength so the constant is not necessary. The constraint equation involves writing the Pythagorean statement for the right triangle formed by d, w, and $2R$ (Fig. 5-4).

The constraint equation $d^2 + w^2 = 4R^2$ can be solved for either d or w and substituted in the defining equation. Either way does not look too appealing.

Solving for w keeps the numbers smaller so write $w = (4R^2 - d^2)^{1/2}$ and substitute into the defining equation to write S in terms of d only.

$$S = (4R^2 - d^2)^{1/2}d^3$$

Differentiate S with the product rule

$$S' = (4R^2 - d^2)^{1/2}(3d^2) + d^3\left(\frac{1}{2}\right)(4R^2 - d^2)^{-1/2}(-2d)$$

$$= 3d^2(4R^2 - d^2)^{1/2} - \frac{d^4}{(4R^2 - d^2)^{1/2}}$$

Set S' equal to zero

$$3d^2(4R^2 - d^2)^{1/2} = \frac{d^4}{(4R^2 - d^2)^{1/2}} \quad \text{or} \quad d^2 = 3(4R^2 - d^2) \quad \text{or} \quad 4d^2 = 12R^2$$

and $d = \pm\sqrt{3}R$. The positive value for d substituted into the constraint equation produces $w^2 = 4R^2 - 3R^2 = R^2$ and $w = R$.

The maximum strength beam that can be cut from a log of radius R is one of dimensions R and $\sqrt{3}R$. It is not necessary to formally determine that this is a maximum. It is the only reasonable choice from the first derivative equals zero condition.

Example 5-5 A park area of 5000 m^2 is to be built in the shape of a rectangle along a river. Fencing will be on three sides. What is the minimum length of fencing for the desired area?

Solution: Fencing is required only on three sides of the rectangle as shown in Fig. 5-5. The defining equation is for the perimeter, the variable we want to minimize: $P = 2a + b$. The constraint equation is from the area requirement. Stated in the form of an equation the constraint is: $ab = 5000$. In order to write P in terms of one variable, solve the area equation for b and substitute.

Fig. 5-5

$$b = \frac{5000}{a} \quad \text{so} \quad P = 2a + \frac{5000}{a} = 2a + 5000a^{-1}$$

Take the derivative of P: $P' = 2 + 5000(-a^{-2})$ and set $P' = 0$:

$$2 = \frac{5000}{a^2} \quad \text{and} \quad a^2 = 2500 \quad \text{or} \quad a = 50.$$

The second derivative of P is $P'' = -5000(-2a^{-3}) = \dfrac{5000}{a^3}$, which is positive for all positive values of a, so $a = 50$ is a minimum. Putting $a = 50$ back into the constraint equation: $50b = 5000$ yields $b = 100$. The dimensions $a = 50$, $b = 100$ provide the minimum fencing requirement. The graph of P vs. a is helpful in understanding this problem. The form $P = 2a + \dfrac{5000}{a}$ is most convenient for graphing. Only positive a has meaning.

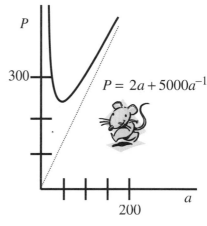

Fig. 5-6

The first step in graphing (see the guidelines for graphing) is to look for dominant terms. There are two here. The $2a$ term dominates for large a and the $5000/a$ term dominates for small a.

In mathematical terms:

$$\text{as } a \to 0, P \to +\infty; \quad \text{and as} \quad a \to \infty, P \approx 2a.$$

The value of the function at $a = 50$, the point where the slope equals zero, is

$$P(50) = 2(50) + 5000/50 = 200.$$

With this information the curve can be sketched as in Fig. 5-6.

Example 5-6   An orange farmer knows from experience that in a certain field 60 orange trees will produce an average of 400 oranges per tree. For each additional tree planted the average yield per tree will drop by 4 oranges. What number of trees will produce maximum total yield?

Solution: The total yield for 60 trees with an average of 400 oranges per tree is:

$$Y|_{60} = (60 \text{ trees}) \frac{400 \text{ oranges}}{\text{tree}} = 24{,}000 \text{ oranges}$$

For one more tree the yield is:

$$Y|_{61} = (61 \text{ trees}) \frac{396 \text{ oranges}}{\text{tree}} = 24{,}156 \text{ oranges}$$

For another tree (total 62) the yield is:

$$Y|_{62} = (62 \text{ trees}) \frac{392 \text{ oranges}}{\text{tree}} = 24{,}304 \text{ oranges}$$

Looking at these numbers, the general formula for total yield as the number of trees is increased is:

$$Y = (60 + x)(400 - 4x)$$

where x is the number of trees in excess of 60.

Problem statements similar to this one can be confusing. You may have already figured that out! One way of getting a handle on the defining equation is to put in some numbers. In this case, writing the total yield for 60 trees producing an average of 400 oranges per tree and then increasing the number of trees by 1 and decreasing the yield per tree by 4, then repeating the process (increasing the number of trees to 62 and decreasing the yield per tree another 4 oranges) provides an education in how to write the general statement for the yield. The numbers also allow you to check the defining equation you have written.

Write the yield equation as $Y = (60 + x)(400 - 4x) = 24{,}000 + 160x - 4x^2$. The first derivative of Y is $Y' = 160 - 8x$ and setting $Y' = 0$, $x = 20$.

The second derivative of Y is $Y'' = -8$ verifying that $x = 20$ is a maximum. The total number of trees for maximum yield is 80 (20 more than the original 60).

Example 5-7 Find the minimum cost to construct a cylindrical container if material for the top and bottom costs 4 cents per square inch and material for the sides costs 3 cents per square inch. The container is to have volume 100 in³.

Solution: Draw a cylinder of radius r and height h. The area of the top and bottom is πr^2. The area of the side is $(2\pi r)h$. Imagine the side as a piece $2\pi r$ long, the circumference of the container, and h high. (See Fig. 5-7.)

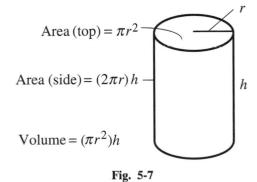

Area (top) $= \pi r^2$

Area (side) $= (2\pi r)h$

Volume $= (\pi r^2)h$

Fig. 5-7

The defining equation is the cost equation which in words is 4 cents times the area of the top and bottom plus 3 cents times the area of the side.

$$C = 4(\pi r^2 + \pi r^2) + 3(2\pi r h) = 8\pi r^2 + 6\pi r h$$

The constraint is that the volume must be 100 in³. The volume of a cylindrical container is the area of the bottom, πr^2, times the height, h: $V = \pi r^2 h$.

Set $V = 100$, solve for h, and substitute into the defining equation: $100 = \pi r^2 h$ or $h = 100/\pi r^2$ and

$$C = 8\pi r^2 + 6\pi r \frac{100}{\pi r^2} = 8\pi r^2 + \frac{600}{r} = 8\pi r^2 + 600r^{-1}$$

The first derivative of C is: $C' = 16\pi r - 600r^{-2}$ and setting $C' = 0$ produces

$$16\pi r - \frac{600}{r^2} = 0 \quad \text{or} \quad r^3 = \frac{600}{16\pi} \quad \text{and} \quad r = \sqrt[3]{\frac{600}{16\pi}} \approx 2.3$$

The second derivative of C is: $C'' = 16\pi + \frac{1200}{r^3}$

C'' is positive for all positive r indicating a minimum for the curve.

Substituting the r for zero slope back into the constraint equation $100 = \pi r^2 h$ produces

$$100 = \pi \left(\frac{600}{16\pi}\right)^{2/3} h \quad \text{or} \quad h = \frac{100}{\pi}\left(\frac{16\pi}{600}\right)^{2/3} \approx 6.1$$

Example 5-8 Postal rates increase when the girth (once around) plus the length of a package exceeds 84 in (Fig. 5-8). What are the dimensions of a "brick-shaped" box with square ends to provide maximum volume?

Solution: The defining equation is the volume, which in this case is the area of the end, x^2, times the length, y: $V = x^2 y$.

The constraint is that the girth, $4x$, plus the length, y, is limited to 84: $4x + y = 84$.

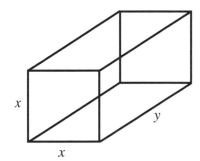

Fig. 5-8

The simplest way to write the $V = \ldots$ equation in one variable is to solve the constraint equation for y: $y = 84 - 4x$ and substitute for y in the defining equation.

$$V = x^2(84 - 4x) = 84x^2 - 4x^3$$

The first derivative of V is $V' = 168x - 12x^2$ and setting $V' = 0$, $12x(14 - x) = 0$ produces two values of x where the slope of V vs. x is zero: $x = 0$ and $x = 14$. The value $x = 0$ produces a zero volume, about as minimal as you can get, so $x = 14$ is a good bet for maximum volume.

The second derivative of V is $V'' = 168 - 24x$. Evaluating V'' at $x = 14$ is $V''(14) = 168 - 24(14) = 168 - 336 = -168$ verifying our suspicion that $x = 14$ produced the maximum volume.

Going back to the constraint equation solved for y, the corresponding y dimension is

$$y = 84 - 4(14) = 84 - 56 = 28.$$

A box with a square end 14 in on a side and length 28 produces the maximum volume within the girth and length restrictions.

It's a Wrap

✔ Write the defining equation

✔ Write the constraint equation

✔ Use the constraint to write the defining in one variable

✔ Differentiate for the max or min

✔ The second derivative confirms max or min

Test Yourself

PROBLEMS

1. Maximize $P = xy$ with the constraint $x + y = 8$.
2. Find the minimum for $G = x^2 + y^2$ with the constraint $2x + y = 8$.
3. Find the maximum and minimum points for the function $f(x) = 2x^3 + 3x^2 - 12x + 1$ and use the second derivative to determine whether each point is a maximum or minimum.
4. Find the maximum and minimum points for the function $y = 4x^3 + 3x^2 - 6x + 1$ and use the second derivative to determine whether the points are maxima or minima.
5. Cylindrical plastic containers are to hold 750 cc. The cost of the cans is directly proportional to the amount of material in the cans. What dimensions (height and radius) produce minimum cost?
6. A rectangular exercise area for your pet rabbit is to be built using an existing L-shaped fence. What dimensions provide minimum fencing and 250 m² area?

7. In a certain manufacturing operation there are at present 40 workers, each worker capable of producing 50 items per day. For each two additional workers production per worker drops by 1 item per day. What is the number of workers for maximum production?

8. Marketing experience indicates that the revenue from the release of a DVD follows $R = 20,000p - 1000p^2$ where p is the price of the DVD. What is the price for maximum revenue?

9. A clothing manufacturer makes sweat shirts. The cost of making x shirts (valid on production runs up to 500) follows:$C(x) = 2000 + 10x + 0.2x^2$. The average cost per shirt is this function divided by the number of shirts produced. $\overline{C(x)} = \dfrac{C(x)}{x} = 2000x^{-1} + 10 + 0.2x$. Find the production run that has the lowest cost per shirt.

10. You need to make open top boxes from an 8×18 in rectangle of material by cutting squares out of the corners and folding up the sides. Design the largest volume box that can be made this way.

ANSWERS

1. Substitute for x in the defining equation: $P = (8 - y)y = 8y - y^2$ and take the derivative. $P' = 8 - 2y$. Set this derivative to zero and find $y = 4$. The second derivative is negative everywhere so this is a maximum. Values of $x = 4$ and $y = 4$ produce maximum P.

2. Substitute for y. $G = x^2 + (8 - 2x)^2 = 5x^2 - 32x + 64$. The first derivative is $G' = 10x - 32$. This gives $x = 3.2$ as the minimum. The second derivative is always positive so this is a minimum.

3. The first derivative is $\dfrac{df}{dx} = 6x^2 + 6x - 12$ which is zero for $x = 1$ and $x = -2$.

   The second derivative is $\dfrac{d^2f}{dx^2} = 12x + 6$ is positive for $x = 1$, indicating a minimum for $x = 1$, and negative for $x = -2$, indicating a maximum, for $x = -2$.

4. The first derivative is $y' = 12x^2 + 6x - 6$. The curve has extrema at $1/2$ and -1. The second derivative is $y'' = 24x + 6$, which is positive at $1/2$ indicating a minimum at this point, and negative at $x = (1$ indicating a maximum at this point.

5. We need to minimize the area. (Area is directly related to cost.) The area of the ends is πr^2 each and the area of the cylinder walls is the circumference of the end caps times the height of the cylinder: $2\pi rh$. The volume is the constraint and the volume is the area of the end caps times the height $\pi r^2 h$. The constraint equation is $750 = \pi r^2 h$, and the defining equation is

$A = 2\pi r^2 + 2\pi rh$. Solve the constraint equation for h, substitute in the equation for area and differentiate.

$$A = 2\pi r^2 + 2\pi r \frac{750}{\pi r^2} = 2\pi r^2 + \frac{1500}{r}$$
$$\frac{dA}{dr} = 4\pi r - \frac{1500}{r^2}$$

Set $\dfrac{dA}{dr} = 0$ so $r^3 = \dfrac{1500}{4\pi} = 119$ and $r = 4.9$

Substitute this r value into the constraint equation to find $h = 9.9$

6. The problem is to minimize the length of fence $a + b$ with the constraint that the area ab be equal to 250.

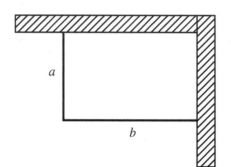

$L = a + b$ with $ab = 250$,
$a = \dfrac{250}{b}$
$L = 250b^{-1} + b$

$\dfrac{dL}{db} = -250b^{-2} + 1$

$\dfrac{250}{b^2} = 1 \quad b^2 = 250 \quad b = 15.8$ and

substituting in $ab = 250$, $a = 15.8$

7. The yield follows the equation $Y = (40 + 2x)(50 - x)$. The first parenthesis represents the number of items produced per worker and the second is the drop off in production for each additional worker. Put in values of 0, 1, and 2 to convince yourself that this is the correct statement of the yield. Write the yield, first derivative to find the extrema in number of workers and second derivative to determine that this number produces maximum yield.

$Y = 2000 + 100x - 40x - 2x^2 = 2000 + 60x - 2x^2$

$Y' = 60 - 4x$ and for $Y' = 0$, $x = 15$

$Y'' = -4$

The second derivative confirms a maximum in yield with the addition of 7.5 workers.

8. The maximum in revenue is the price when d revenue/d price is equal to zero and the second derivative is negative.

$R = 20,000p - 1000p^2$
$R' = 20,000 - 2000p$

$R' = 0$ when $p = 10$

$R'' = -2000$ maximum

Maximum revenue is for a price of $10.00 for the DVD.

9. The problem asks for the minimum in the $\overline{C(x)}$ curve, so find the first derivative.

$\overline{C(x)}' = -2000x^{-2} + 0.2$ and setting this equal to zero

$$\frac{2000}{x^2} = 0.2 \quad \text{so} \quad x = 100$$

The second derivative is positive everywhere so a production run of 100 shirts produces the minimum cost per shirt.

10. Draw an 8×18 in rectangle showing squares at the corners. The volume, the quantity to be maximized is $V = a(18 - 2a)(8 - 2a)$. Take the derivative of V and find the value of a for maximum volume.

$V = 144a - 52a^2 + 4a^3$

$V' = 144 - 104a + 12a^2$

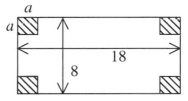

Solving for $V' = 0$ by quadratic formula produces two values, 1.7 and 6.9. The second derivative is $V'' = -104 + 24a$ and is negative for 1.7 and positive for 6.9 indicating that the 1.7 value is the maximum value. In addition, the only value of a that will work is 1.7. An a value of 6.9 makes the cut out squares greater than the width of the piece. For $a = 1.7$, the volume is 114.

CHAPTER 6

RELATED RATE PROBLEMS

Do I Need to Read This Chapter?

You should read this chapter if you need to review or you need to learn about

➡ Find related rates through implicit differentiation

➡ Solve distance-related rates

➡ Related rates for moving bodies

➡ Geometric-related rates

➡ Area-related rates

➡ Volume-related rates

Related rate problems relate one rate, written as a derivative, to another rate written as a derivative. An excellent example of a related rate problem, and one that is in nearly every calculus book including this one, is a ladder sliding down a wall. (See Fig. 6-1.) The top of the ladder is moving down the wall while the bottom of the ladder is moving away from the wall. The rate (speed) the top is moving down the wall can be related to the rate (speed) the bottom is moving away from the wall. This problem illustrates well the name of these problems; related rate problems.

A little review is in order. Related rate problems are similar to problems involving implicit differentiation. Equations in the form $y = f(x)$ such as $y = x^2 + 2x - 3$ are differentiated term by term according to the rules for differentiating polynomials, products, quotients, or whatever. Equations where the x's and y's are mixed together so the equation cannot be written as $y = f(x)$ or $x = f(y)$ (an x alone or a y alone on one side of the equation) are differentiated implicitly.

For example, the equation $2xy^2 + xy^3 = 0$ must be differentiated implicitly as

$$2y^2dx + 4xydy + 3xy^2dy + y^3dx = 0$$

with dy/dx formed by grouping and rearranging.

If x and y could both change over time then a related rate associated differentiation of this equation would be

$$2y^2\frac{dx}{dt} + 4xy\frac{dy}{dt} + 3xy^2\frac{dy}{dt} + y^3\frac{dx}{dt} = 0$$

In this statement, $\dfrac{dx}{dt}$ is directly related to $\dfrac{dy}{dt}$

$$(2y^2 + y^3)\frac{dx}{dt} = -(4xy + 3xy^2)\frac{dy}{dt} \quad \text{or} \quad \frac{dx}{dt} = -\frac{4xy + 3xy^2}{2y^2 + y^3}\frac{dy}{dt}$$

This is an example of a related rate differentiation. Now take a look at perhaps not the simplest related rate problem, but possibly the simplest to visualize. Notice how this problem is written. The general situation is described, then a rate is specified and the related rate is requested for a certain condition.

Example 6-1   A 7-m-long ladder is sliding down a wall. The bottom of the ladder is pulled from the wall at 1.5 m/s. What is the rate at which the top of the ladder is going down when the bottom is 3 m from the wall?

Solution: Help to visualize the problem by sketching a ladder leaning against a wall with the bottom being pulled out from the wall at $dx/dt = 1.5$ m/s (Fig. 6-1).

The question, written in mathematical language, is: "What is dy/dt when $x = 3$ m and $dx/dt = 1.5$ m/s?"

In max-min problems the defining equation is a mathematical statement of the problem. In related rate problems the defining equation is sometimes a little more obscure, actually sometimes a lot more obscure! Look at the ladder in the graphic and think of a way to relate x to y. Don't start by trying to write the dx/dt and dy/dt. The rates come out of the differentiation.

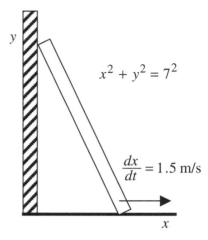

$$x^2 + y^2 = 7^2$$

$$\frac{dx}{dt} = 1.5 \text{ m/s}$$

Fig. 6-1

Quick Tip

The hardest part of related rate problems is to see, and then write down, a relationship between the variables. Writing this defining equation that ties the variables together is the key step in related rate problems.

In this problem the Pythagorean theorem for a right triangle relates x and y.

The defining equation is $x^2 + y^2 = 7^2$ and taking $\frac{d}{dt}$ we write

$$2x\frac{dx}{dt} + 2y\frac{dy}{dt} = 0 \quad \text{or} \quad \frac{dy}{dt} = -\frac{x}{y}\frac{dx}{dt}$$

Now the numbers can be put in the equation to find dy/dt when $dx/dt = 1.5$ m/s and $x = 3$ m. What about the y in the denominator? The y can be determined from the Pythagorean relation $y = \sqrt{7^2 - 3^2} \approx 6.3$. With these numbers, dy/dt is calculated as

$$\frac{dy}{dt} = -\frac{x}{y}\frac{dx}{dt} = -\frac{3}{6.3}(1.5 \text{ m/s}) = -0.71 \text{ m/s}$$

The top of the ladder is coming down the wall at 0.71 m/s when the bottom of the ladder is 3 m away from the wall and moving at 1.5 m/s.

Example 6-2 A girl is flying a kite. The kite is moving horizontally at a height of 120 ft when 250 ft of string is out and the rate of increase in string length is 2 ft/s. How fast is the kite moving in the *x*-direction for these conditions?

Solution: Visualize the problem and set up a right triangle with the height, horizontal direction, and string. In this problem the kite only moves horizontally, and the string is straight—idealized—but the conditions make for a problem that can be solved.

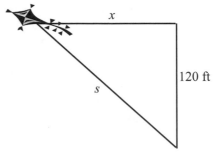

Fig. 6-2

Referring to Fig. 6-2, the problem question can be written in mathematical terms as:

What is $\dfrac{dx}{dt}$, when $\dfrac{ds}{dt} = 2$, the height of the kite is 120 ft, and the distance out is 250 ft?

Go back over the problem statement and practice changing the problem statement into this mathematical statement. One of the more challenging parts of any calculus problem is translating the words into mathematical statements.

The Pythagorean theorem relates the variables *x* and *s* in the right triangle:

$$120^2 + x^2 = s^2.$$

Take $\dfrac{d}{dt}$ to get $2x\dfrac{dx}{dt} = 2s\dfrac{ds}{dt}$ or $\dfrac{dx}{dt} = \dfrac{s}{x}\dfrac{ds}{dt}$ The *ds/dt* rate (2 ft/s) is given in the problem as is the height (120 ft) and the distance out (250 ft). The *x* value for these conditions can be calculated from the Pythagorean theorem:

$$x^2 = s^2 - 120^2 \quad \text{or} \quad x = \sqrt{250^2 - 120^2} = 219$$

Now the numbers can be put into the formula for *dx/dt*:

$$\frac{dx}{dt} = \frac{s}{x}\frac{ds}{dt} = \frac{250\,\text{ft}}{219\,\text{ft}}\,(2\,\text{ft/s}) = 2.3\,\text{ft/s}$$

When the kite is 250 ft away from the girl, at a height of 120 ft, and the string is going out at 2 ft/s, the kite is moving 2.3 ft/s horizontally.

These first two problems have utilized the Pythagorean theorem as their defining or "getting started" equation. Related rate problems use a variety of defining statements to tie the variables together. As you go through this chapter be
aware of the various techniques for relating the variables. If you see a related
rate problem on a test that can be analyzed with the Pythagorean theorem, you
will know how to do that problem.

This next problem uses the Pythagorean theorem but it has another little twist.
The information for the problem is given primarily in terms of rates, and the
solution involves three different rates.

Example 6-3   Two ships are traveling at right angles. The first ship, traveling
at 8 m/s, crosses the path of the second ship when it is 1000 m away (from the
point where the paths cross) and traveling at 6 m/s. What are their positions,
separation, and rate of separation 300 s after their paths cross?

Solution: Diagram the problem on an *x-y* coordinate system with the first ship
going in the *y*-direction and the second ship going in the *x* direction. Figure 6-3
is for *t* = 0, the time when the ships cross paths. The drawing helps to visualize
the problem.

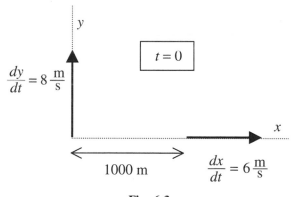

Fig. 6-3

The position of the first ship at any time *t* is *y* = (8 m/s)*t*. The position of the
second ship at any time *t* is *x* = 1000 m + (6 m/s)*t*. The separation of the ships
is from the Pythagorean theorem $s = \sqrt{x^2 + y^2}$.

The position of the first ship at 300 s is its speed (8 m/s) times the 300 s

$$y|_{300} = (8 \text{ m/s})(300 \text{ s}) = 2400 \text{ m}$$

The position of the second ship at 300 s is the 1000 m plus the 6 m/s times the 300 s

$$x|_{300} = 1000 \text{ m} + (6 \text{ m/s})(300 \text{ s}) = (1000 + 1800) \text{ m} = 2800 \text{ m}$$

The separation of the ships is a straight Pythagorean theorem problem.

$$s = \sqrt{2400^2 + 2800^2} = 3688 \text{ m}$$

The rate at which they are separating is the fun, that is to say calculus, part of the problem. The rate at which they are separating is, in calculus talk, ds/dt, and we already have the dx/dt and dy/dt. Start with the separation written in Pythagorean theorem form $s = (x^2 + y^2)^{1/2}$ and differentiate, carefully.

Writing

$$ds = \frac{1}{2}(x^2 + y^2)^{-1/2}d(x^2 + y^2)$$

as the first step will help to prevent errors with (1/2)s and the minus signs.

Continuing, $ds = \frac{1}{2}(x^2 + y^2)^{-1/2}(2xdx + 2ydy)$ and finally

$$\frac{ds}{dt} = \frac{1}{(x^2 + y^2)^{1/2}}\left(x\frac{dx}{dt} + y\frac{dy}{dt}\right)$$

This rate of separation is to be evaluated at $t = 300$ s (Fig. 6-4).

$$\frac{ds}{dt}\bigg|_{300} = \frac{1}{3688 \text{ m}}[(2800 \text{ m})(6 \text{ m/s}) + (2400 \text{ m})(8 \text{ m/s})] = 9.8\frac{\text{m}}{\text{s}}$$

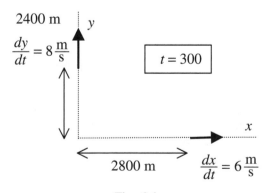

Fig. 6-4

Example 6-4   A 3-ft tall penguin (Penny) is taking a leisurely stroll at 0.5 ft/s
away from a 12-ft tall penguin way light. What is the length of her shadow
and how fast is the tip of her shadow moving when she is 40 ft away from the
light?

Solution: When you see a triangle in a related rate problem look for similar
triangles. Don't start the problem looking for derivatives. Concentrate on the
defining equation for the problem. The derivatives come later.

Your first order of business in a related rate problem is to find relationships
between the variables. In this problem set up the triangle, complete with known
numbers, and then label some of the distances. The change in length of the
hypotenuse of this triangle is not what we are looking for. It is lengths along the
ground: the length from the light to Penny and the length of her shadow. Take x
as the length from beneath the light to Penny, and z as the length from beneath
the light to the end of her shadow. The length of her shadow is $z - x$. Draw this
triangle (refer to Figs. 6-5 and 6-6).

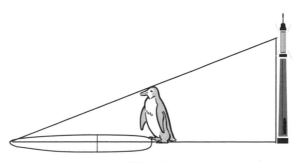

Fig. 6-5

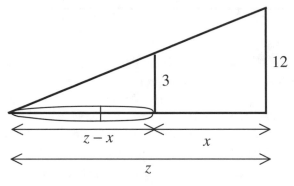

Fig. 6-6

Notice that the triangle with sides $z - x$ and 3 is similar to the triangle with sides z and 12. Similar triangles are triangles with the same angles and their sides in proportion. This means that the ratios of the sides are equal.

$$\frac{z - x}{3} = \frac{z}{12}$$

Eliminating the fraction, $12z - 12x = 3z$ or $9z = 12x$ or $3z = 4x$, produces a simple relationship between x and z. The related derivative rates are

$$\frac{dz}{dt} = \frac{4}{3}\frac{dx}{dt}$$

Notice that x and z don't enter into the rate relationship. Penny is moving at

$$\frac{dx}{dt} = 0.5\frac{\text{ft}}{\text{s}}$$

so the tip of her shadow is moving at

$$\frac{dz}{dt} = \frac{4}{3}\frac{dx}{dt} = \frac{4}{3}\left(0.5\frac{\text{ft}}{\text{s}}\right) = 0.67\frac{\text{ft}}{\text{s}}$$

Since Penny is walking away from the light at 0.5 ft/s and the tip of her shadow is growing at 0.67 ft/s her shadow is getting longer as she moves away from the light.

As an exercise, go to a desk or table with a lamp. Place a pencil near the lamp and observe the length of the shadow. The pencil should be a foot or so from the light and perhaps slightly tilted. Now move the pencil away from the light and observe the shadow. The shadow will grow and the tip of the shadow will move faster than the pencil.

Example 6-5 A conical container of base radius 5 ft and height 10 ft is being filled with sand at the rate of 2 ft³/min. How fast is the level of the sand rising when it is 6 ft above the apex of the conical container? (See Fig. 6-7.)

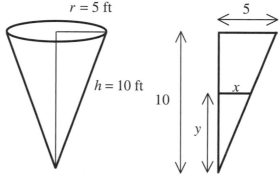

$r = 5$ ft

$h = 10$ ft

5

10

x

y

Fig. 6-7

Solution: The formula for the volume of a cone is, from the Mathematical Tables in the back of the book, $V = (1/3)\pi r^2 h$.

The dimensions defining the cone are given in the problem so calculating the total volume of the container is not a problem.

Sketch the cone, and next to the cone sketch the profile of the entire cone and a partially filled cone with radius x and height y. This is another similar triangles problem! The radius to height ratio is the same for any radius and depth. In this case the similar triangles are the ones with sides x and y, and 5 and 10.

The similar triangle statement is $\frac{5}{10} = \frac{x}{y}$ or $x = \frac{y}{2}$. The question "How fast is the level of the sand rising . . .?" means, what is dy/dt? Knowing dV/dt and requiring dy/dt, we need to write V in terms of y only. Time derivatives of V in terms of y will produce a relation between dV/dt and dy/dt.

Substitute in the V equation:

$$V = \frac{1}{3}\pi x^2 y = \frac{1}{3}\pi\left(\frac{y}{2}\right)^2 y = \frac{\pi}{12}y^3.$$

And taking derivatives produces

$$\frac{dV}{dt} = \frac{\pi}{4}y^2\frac{dy}{dt} \quad \text{or} \quad \frac{dy}{dt} = \frac{4}{\pi y^2}\frac{dV}{dt}.$$

Adding numbers for $y = 6\,\text{ft}$,

$$\frac{dy}{dt} = \frac{4}{\pi y^2}\frac{dV}{dt} = \frac{4}{(36\,\text{ft}^2)\pi}\left(2\frac{\text{ft}^3}{\text{min}}\right) = 0.071\frac{\text{ft}}{\text{min}}.$$

At a depth of 6 ft the sand is rising at 0.071 ft/min.

Another category of related rate problems involves increasing or decreasing area, volume, radius of a sphere, or some other geometric property. These next two problems involve geometry. In general, geometry problems are not overly difficult, usually involving just one equation.

Example 6-6   A circular oil slick is forming in such a way that the radius of the slick is increasing at a constant rate of 12 ft/hr. What will be the rate of area increase when the slick has radius 300 ft? (See Fig. 6-8.)

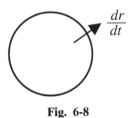

Fig. 6-8

Solution: The area is related to the radius by $A = \pi r^2$ (see the Mathematical Tables). The rate of A and the rate of r are directly available from this one equation.

$$\frac{dA}{dt} = 2\pi r\frac{dr}{dt}$$

Using the numbers given in the problem

$$\left.\frac{dA}{dt}\right|_{300} = 2\pi r\frac{dr}{dt} = 2\pi\,(300\,\text{ft})\left(12\frac{\text{ft}}{\text{hr}}\right) = 22{,}600\,\frac{\text{ft}^2}{\text{hr}}$$

The area of the oil slick is increasing at 22,600 ft²/hr when the radius is 300 ft.

Example 6-7   The deployment of the safety air bags in automobiles is very much like blowing up a balloon. If the air bag expands under a volume change of 10,000 cm³/s, what is the radial change when the bag has 5 cm radius and 10 cm radius? What is the significance of your result?

Solution: The volume of the bag is

$$V = \frac{4}{3}\pi r^3$$

The rate relations are

$$\frac{dV}{dt} = \frac{4}{3}\pi\left(3r^2\frac{dr}{dt}\right) \quad \Rightarrow \quad \frac{dr}{dt} = \frac{1}{4\pi r^2}\frac{dV}{dt}$$

Now put in the numbers for the 5 cm and 10 cm radii.

$$\left.\frac{dr}{dt}\right|_5 = \frac{1}{4\pi(5\text{ cm})^2}10{,}000\frac{\text{cm}^3}{\text{s}} = 32\frac{\text{cm}}{\text{s}}$$

$$\left.\frac{dr}{dt}\right|_{10} = \frac{1}{4\pi(10\text{ cm})^2}10{,}000\frac{\text{cm}^3}{\text{s}} = 8\frac{\text{cm}}{\text{s}}$$

Doubling your distance from the air bag decreases its speed when it hits you by one-fourth. Conclusion: Sit as far away from the air bag as you can!

Example 6-8 An obstruction in an artery is to be removed by inflating a spherical balloon in the artery. The rate of increase of the radius of the balloon must be limited to 1 mm/min when the radius is 4 mm. What is the maximum volume rate increase, the rate at which oxygen is pumped into the balloon, corresponding to this radius rate increase? (See Fig. 6-9.)

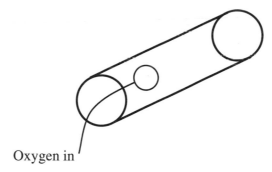

Oxygen in

Fig. 6-9

Solution: The volume of a sphere is $V = (4/3)\pi r^3$ (see the Mathematical Tables).

Again, the rate relations are immediately available from this equation for the volume of a sphere.

$$\frac{dV}{dt} = 4\pi r^2 \frac{dr}{dt}$$

Evaluating at $r = 4$ mm and $\dfrac{dr}{dt} = 1\ \dfrac{\text{mm}}{\text{min}}$

$$\left.\frac{dV}{dt}\right|_4 = 4\pi(4\,\text{mm})^2\left(1\frac{\text{mm}}{\text{min}}\right) = 201\,\frac{\text{mm}^3}{\text{min}}$$

The maximum rate that the balloon can be filled at the 4 mm radius is 201 mm³/min.

These last two problems are illustrative of problems where the formulas are given to you. In most of these types of problems, differentiating the formula is the challenge.

Example 6-9 When the price of a certain product is p dollars per unit, customer demand is x hundreds of units (per month). The relation between p and x is $x^2 + 2px + 0.5p^2 = 80$. When the price is \$4.00 and dropping at the rate of \$0.25 per month, what is the rate of increase in demand?

Solution: This equation requires an implicit type of differentiation to find dp/dt, the rate of price change, and dx/dt, the rate of demand change.

$$2x\frac{dx}{dt} + 2p\frac{dx}{dt} + 2x\frac{dp}{dt} + p\frac{dp}{dt} = 0 \quad \text{or} \quad (2x + 2p)\frac{dx}{dt} + (2x + p)\frac{dp}{dt} = 0$$

or
$$\frac{dx}{dt} = -\frac{2x + p}{2x + 2p}\frac{dp}{dt}$$

The rate of price change, dp/dt, is given in the problem as is p, the price. The demand rate, x, is not given and must be computed from the original equation. Substituting for $p = 4$ ($p = \$4.00$) in $x^2 + 2px + 0.5p^2 = 80$ yields $x^2 + 8x + 8 = 80$ or $x^2 + 8x - 72 = 0$. The quadratic formula produces two answers. The positive 5.4 is the realistic one.

$$x = \frac{-8 \pm \sqrt{64 - 4(1)(-72)}}{2(1)} = \frac{-8 \pm 18.8}{2} = 5.4, -14.4$$

With all the needed values, dx/dt can be evaluated. Watch the signs closely.

$$\left.\frac{dx}{dt}\right|_4 = -\frac{2(5.4) + 4}{2(5.4) + 8}(-0.25) = \frac{14.8}{18.8}(0.25) = 0.20$$

The demand rate is increasing by 0.20 hundreds of units per month when the price is $4.00 and dropping at the rate of $0.25 per month. Carrying the units through this problem is difficult because the constants in the original equation must have the appropriate units to make each term in the equation have the same units.

Example 6-10 The amount of trash, measured in thousands of pounds, accumulating in a city dump follows the formula $T = 1.3p^2 - 100p + 30$, where p is the population in hundreds of thousands. What is the rate of trash increase when the population is 200 thousand and increasing by 0.2 thousand (0.1%) per month?

Solution: Relating the rate of trash increase, dT/dt, to the population increase, dp/dt, comes directly from implicit type differentiation of the expression for the amount of trash.

$$\frac{dT}{dt} = 2.6p\frac{dp}{dt} - 100\frac{dp}{dt} = (2.6p - 100)\frac{dp}{dt}$$

The population and the rate of increase in population are given in the problem so we have

$$\left.\frac{dT}{dt}\right|_{200} = [(2.6)(200) - 100][0.2 \text{ thousand per month}]$$

$$= 84 \text{ thousand of pounds per month.}$$

This is also an interesting max-min problem. Take $\dfrac{dT}{dp} = (2.6p - 100)$ and set equal to zero to find $p = 38$. The second derivative $\dfrac{d^2T}{dp^2} = 2.6$ so the point $p = 38$ is a minimum.

The minimum in trash accumulation rate is at 38 thousand people, but at 200 thousand the curve becomes progressively more positive and the trash problem progressively worse.

Example 6-11 The number of CD players that can be manufactured by a certain company follows a Cobb-Douglas type of production model, where q, the production output, depends on the number of workers, n, and the number of automatic assembly machines, r, according to $q = 20n^{0.6}r^{0.4}$. With 20 workers the company is producing 300 CDs per day and has sufficient revenue to purchase one automatic assembly machine per month. How many workers per month should be laid off to maintain constant production?

Solution: The first step in the problem is to perform an implicit differentiation remembering that q has to remain constant.

$$0 = 20[n^{0.6}(0.4)r^{-0.6}dr + r^{0.4}(0.6)n^{-0.4}dn]$$

Now write the rate relationship. This could have been written directly.

$$0.4\left(\frac{n}{r}\right)^{0.6}\frac{dr}{dt} = -0.6\left(\frac{r}{n}\right)^{0.4}\frac{dn}{dt}$$

$$\frac{dn}{dt} = -\frac{2}{3}\left(\frac{n}{r}\right)\frac{dr}{dt}$$

It takes 20 workers to produce 300 CDs so the number of automatic assembly machines can be calculated from the original equation.

$$q = 20n^{0.6}r^{0.4} \quad \Rightarrow \quad 300 = (20)(20)^{0.6}r^{0.4}$$

$$r^{0.4} = \frac{300}{(20)(20)^{0.6}} = \frac{15}{20^{0.6}}$$

$$[r^{0.4}]^{2.5} = r = \left[\frac{15}{(20)^{0.6}}\right]^{2.5} = \frac{(15)^{2.5}}{(20)^{1.5}} = 9.7$$

Now calculate the related rate.

$$\frac{dn}{dt} = -\frac{2}{3}\left(\frac{20}{9.7}\right)(1) = -1.4$$

Adding one automatic assembly machine and releasing 1.4 workers per month will maintain production.

It's a Wrap

✔ Relate the variables in the problem

✔ Use a diagram if that will help

✔ Use a variation of implicit differentiation to write the rates

✔ Calculate specific variable values from the defining equation

Test Yourself

PROBLEMS

1. For the equation $x^2 + y^2 = 20$ find dy/dt when $x = 4$ and $dx/dt = -6$.
2. For $x^2 - 2xy + y^2 = 0$ find dx/dt when $y = 2$ and $dy/dt = 3$.

3. A rocket launched vertically is tracked by a camera 200 m horizontally away from the launch point. What is the vertical speed of the rocket when it is 600 m from the camera with this distance (from the camera) increasing at 80 m/s?

4. The cost in dollars per day to produce thumb drives follows $C(x) = 4000 - 12x + \dfrac{x^2}{200}$ where x is the number of thumb drives produced each day. The present production level is 200 thumb drives per day and this rate is increasing at 2 per day. How fast is the average cost changing? Remember that average cost is $C(x)$ over x.

5. At a price, p, a manufacturer is willing to supply x number of products according to the equation $x^2 - xp - p^2 = 30$. What is the rate of supply of x when p is $9.00 and increasing at $0.20 per week?

6. The mass of a raindrop is $M = \rho\dfrac{4}{3}\pi r^3$ where ρ is the density of water, 1 g/cm³, and r is the radius of the drop. At what rate is the radius of a raindrop changing when it has a radius of 10 mm and is gaining water at the rate of 0.01 g/min?

7. An observer with a telescope is observing the approach of an airplane traveling 700 km/hr towards a point directly over the telescope. The airplane remains at a constant height of 9.0 km. What is the angular rate of change (in rad/sec) for the telescope when the airplane is 24 km horizontally from the telescope?

8. An adiabatic process follows the pressure-volume formula $PV^{1.4} = C$, a constant. For $P = 2$ N/m², $V = 0.8$ m³ and V is increasing at 0.6 m³/s, what is happening to P.

9. The period, the time for one back-and-forth excursion, of a pendulum is $T = 2\pi\sqrt{\dfrac{L}{g}}$, where L is the length of the pendulum and g is the accelera tion due to gravity equal to 9.8 m/s². If the length of the pendulum stretches by 0.01% over 1 year, what is the change in the period?

10. The Poiseuille's law for flow through a cylindrical pipe (blood vessel) is: $v = \dfrac{K}{L}(R^2 - r^2)$, where v is the velocity, K is a constant equal to 6 1/s, L is the length of the pipe, R is the radius of the pipe and r is the distance out from the center line of the pipe. For a 100-cm pipe of radius 0.2 cm, find the rate that the flow velocity is changing half-way between the center line and the wall when the pipe is contracting at a rate of 0.0004 cm/s?

ANSWERS

1. Start with the implicit derivative $2xdx + 2ydy = 0$ and proceed to time derivatives $y\dfrac{dy}{dt} = -x\dfrac{dx}{dt}$ and to $\dfrac{dy}{dt} = -\dfrac{x}{y}\dfrac{dx}{dt}$. Now stop and calculate y for $x = 4$. $4^2 + y^2 = 20 \Rightarrow y = 2 \Rightarrow \dfrac{dy}{dt} = -\dfrac{4}{2}(-6) = 12$.

2. $2x\dfrac{dx}{dt} - 2x\dfrac{dy}{dt} - 2y\dfrac{dx}{dt} + 2y\dfrac{dy}{dt} = 0 \Rightarrow (x - y)\dfrac{dx}{dt} = (x - y)\dfrac{dy}{dt}$ Whatever

the values for x and y are, $\dfrac{dx}{dt} = \dfrac{dy}{dt}$, so $\dfrac{dx}{dt} = 3$.

3. Use a diagram. This is a Pythagorean theorem problem where the variables are from the statement: $x^2 + y^2 = L^2$. In this problem the x is constant so take the rate derivatives. $2y\dfrac{dy}{dt} = 2L\dfrac{dL}{dt} \Rightarrow \dfrac{dy}{dt} = \dfrac{L}{y}\dfrac{dL}{dt}$
Calculate y when L is 600 m.
$200^2 + y^2 = 600^2 \Rightarrow y = 566$ Now calculate the vertical speed of the rocket.
$\dfrac{dy}{dt} = \dfrac{600}{566}80$ m/s $= 85$ m/s

4. The average cost is $\overline{C}(x) = \dfrac{C(x)}{x} = 4000x^{-1} - 12 + \dfrac{x}{100}$. The rate statement is $\dfrac{d\overline{C}(x)}{dt} = (4000)(-1)x^{-2}\dfrac{dx}{dt} + \dfrac{1}{100}\dfrac{dx}{dt} = \left[\dfrac{1}{100} - \dfrac{4000}{x^2}\right]\dfrac{dx}{dt} = \left[\dfrac{1}{100} - \dfrac{4000}{200^2}\right](2) = -0.18$

At this point the average cost of producing the thumb drives is dropping at the rate of \$0.18 per day.

5. Write the rate statement: $2x\dfrac{dx}{dt} - x\dfrac{dp}{dt} - p\dfrac{dx}{dt} - 2p\dfrac{dp}{dt} = 0$. Want to know $\dfrac{dx}{dt}$ when $p = 9$ and $\dfrac{dp}{dt} = 0.20$ per week.

$(2x - p)\dfrac{dx}{dt} = (2p + x)\dfrac{dp}{dt} \Rightarrow \dfrac{dx}{dt} = \dfrac{2p + x}{2x - p}\dfrac{dp}{dt}$

Now find x when $p = 9$. $x^2 - 9x - 81 = 30$ This equation is solved by quadratic formula to yield $x = 16$, so

$\dfrac{dx}{dt} = \dfrac{18 + 16}{32 - 9}\dfrac{dp}{dt} = 1.5(0.20) = 0.30$

The manufacturer is willing to supply 0.30 more items per week or roughly 1 more every 3 weeks for these conditions.

6. $\dfrac{dM}{dt} = \rho\dfrac{4}{3}\pi(3)r^2\dfrac{dr}{dt} = 4\rho\pi r^2\dfrac{dr}{dt}$

$\dfrac{dr}{dt} = \dfrac{1}{4(1\,\text{g/cm}^3)\pi(1\,\text{cm})^2}\left(0.01\dfrac{\text{g}}{\text{min}}\right) = 0.0025\dfrac{\text{cm}}{\text{min}}$

7. This requires a drawing. Referring to the diagram, $\dfrac{dx}{dt} = -700$ km/hr and $y = 9.0$ km. To relate the angle use the tangent function, $\tan\theta = \dfrac{9}{x}$ or $9 = x\tan\theta$. Take the rate derivatives.

$$x\sec^2\theta\,\dfrac{d\theta}{dt} \;+\; \tan\theta\,\dfrac{dx}{dt} = 0$$

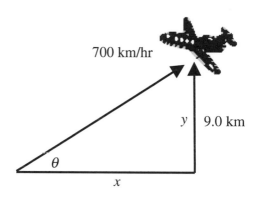

700 km/hr

y | 9.0 km

θ

x

When x is 24, the angle is 21°, so $24(\sec^2 21°)\dfrac{d\theta}{dt} = -\tan 21°\left(-700\,\dfrac{\text{km}}{\text{hr}}\right)$

$$\dfrac{d\theta}{dt} = \dfrac{(\cos^2 21°)\tan 21°}{24\ \text{km}}(700)\,\dfrac{\text{km}}{\text{hr}} = 9.8\,\dfrac{1}{\text{hr}}\,\dfrac{\text{hr}}{60\ \text{min}} = 0.16\,\dfrac{\text{rad}}{\text{min}}$$

This is roughly 0.3° per min.

8. Take the rate derivatives:

$$P(1.4)V^{0.4}\dfrac{dV}{dt} + V^{1.4}\dfrac{dP}{dt} = 0 \quad \dfrac{dP}{dt} = -\dfrac{(1.4)PV^{0.4}}{V^{1.4}}\dfrac{dV}{dt} \quad \dfrac{dP}{dt} = -\dfrac{(1.4)P}{V}\dfrac{dV}{dt}$$

Put in the numbers.

$$\dfrac{dP}{dt} = -\dfrac{(1.4)(2\ \text{N/m}^2)}{(0.8\ \text{m}^3)}[0.6\ \text{m}^3/\text{s})] = 2.1\dfrac{\text{N/m}^2}{\text{s}}$$

9. Start with $T = \dfrac{2\pi}{\sqrt{g}}L^{1/2}$ and take the rate derivative. $\dfrac{dT}{dt} = \dfrac{2\pi}{\sqrt{g}}\dfrac{1}{2}L^{-1/2}\dfrac{dL}{dt}.$

The $\dfrac{dL}{dt} = 0.0001\,L.$ Put this into the rate equation. $\dfrac{dT}{dt} =$

$\dfrac{1}{2}\dfrac{2\pi}{\sqrt{g}}\dfrac{1}{\sqrt{L}}(0.0001)L = \dfrac{0.0001}{2}2\pi\sqrt{\dfrac{L}{g}}.$ Notice that this can be written in terms of the original period. $\frac{dT}{dt} = 0.00005\,T$. Writing this another way is instructive. $\frac{dT}{T} = 0.00005dt$. This states that the fractional change in the period is 0.00005 per year.

10. Write the general rate statement. $\frac{dv}{dt} = \frac{K}{L}(-)2r\frac{dr}{dt}$ Now put in the numbers.

$$\frac{dv}{dt} = -2\frac{6/\mathrm{s}}{1.0\;\mathrm{m}}\;(0.002\;\mathrm{m})\left(0.0004\;\frac{\mathrm{cm}}{\mathrm{s}}\right) = -1.0 \times 10^{-5}\;\frac{\mathrm{cm}}{\mathrm{s}^2}.$$

CHAPTER 7

INTEGRATION

Do I Need to Read This Chapter?

You should read this chapter if you need to review or you need to learn about

➡ The antiderivative

➡ Area under the curve

➡ Average value of a function

➡ Area between curves

There are many calculus problems where the derivative of a function is known and the function is desired. For example, if a mathematical expression for the rate of population growth dP/dt is known, is it possible to "work backwards" to find the expression for P, how the population varies over time?

The process of starting with a derivative and working back to the function is quite naturally called the antiderivative. The antiderivative of a function is an easy concept but often is operationally difficult. There are many integration problems where finding the antiderivative will prove a major challenge.

In some problems the integral can be viewed as the area under the curve of the function being integrated. This is often very helpful in getting a physical "feel" for the problem and the process of integration. This view of the integral will be discussed later in the chapter.

Some problems in integration require a great deal of imaginative thinking and manipulative ability. The simplest first approach to integration is via the antiderivative. After that we will move on to using the area under the curve approach and finally to the more difficult integral problems.

7-1   The Antiderivative

Start with a simple function, $y = x^2$. The derivative of that function is written as $dy = 2xdx$ and finally as $dy/dx = 2x$.

Keeping this short review of differentiation in mind, suppose we encountered a derivative

$$\frac{du}{dv} = 2v$$

and want to know how u varies with v. Keep the differential (of $y = x^2$) in front of you and just work backwards

$$\frac{du}{dv} = 2v \text{ can be written as } du = 2vdv$$

Now all we need to do is perform the inverse or "anti" derivative operation to find u in terms of v. This being mathematics, no operation can be performed without a symbol. For integration we use this elongated "s" shape, so write

$$\int du = \int 2vdv$$

The left side of this equation is the integral of the differential, two inverse operations. The d acting on u is the derivative, while the $\int$ acting on du is the antiderivative. The result of these inverse operations on u is that the left side of this equation is u. The operation is somewhat like squaring a square root. The right side is not so easy except that we have the differential example just above us. The differential of x^2 is $2x\,dx$, so the integral of $2v\,dv$ is v^2. The function described by the differential statement $du = 2v\,dv$ is, therefore, $u = v^2$.

Conceptually, the antiderivative is not difficult. Actually, finding the antiderivative of a complicated function is often not at all easy. Polynomials are the easiest to work with and that is where we will start.

Example 7-1   Find $\int y^3\,dy$.

Solution: We seek a function that differentiates to $y^3\,dy$. The differential of y^4 is $4y^3\,dy$, which is very close to what we want. The differential of $y^4/4$ is $y^3\,dy$ so the $\int y^3\,dy$ is $y^4/4$.

Check the answer by differentiating it. The differential of $\int y^3\,dy$ is $y^3\,dy$(inverse operations), and the differential of $y^4/4$ is $y^3\,dy$. Doing a few integrals of polynomials leads to a general formula for integrating polynomials.

Remember

$$\int x^n\,dx = \frac{x^{n+1}}{n+1}$$

This formula is valid for all n, including fractions and negative exponents, except $n = -1$.

That special case will be taken up later and in more detail in Chapter 8, Exponents and Logarithms. With this general formula for integrating polynomials take the integrals of some other differentials.

Example 7-2   Find the function x in terms of t, starting with the differential statement

$$\frac{dx}{dt} = \frac{1}{3}t^3 + 5t^2 + 4$$

Solution: First rewrite the problem as $dx = \left(\frac{1}{3}t^3 + 5t^2 + 4\right)dt$.

The integral of dx is x so write $x = \int \left(\frac{1}{3}t^3 + 5t^2 + 4\right)dt$.

Most formal integral problems are presented in this form. Now perform the integration term by term, the same way the differential was formed to produce this integral

$$x = \frac{1}{3}\left(\frac{t^4}{4}\right) + 5\left(\frac{t^3}{3}\right) + 4t = \frac{t^4}{12} + \frac{5t^3}{3} + 4t$$

Don't forget that the integral of a constant times dt is the constant times t.

The antiderivative as described so far is not the complete story of antiderivatives, as is illustrated in the next problem. Take a look at a simple function, $y = x^2 + 2x + 7$. The derivative is $dy/dx = 2x + 2$. Now take the antiderivative of $2x + 2$.

$$dy = (2x + 2)dx \quad \text{and} \quad y = \int (2x + 2)dx = 2\frac{x^2}{2} + 2x = x^2 + 2x$$

Where did the 7 in the original function go? Differentiating the function produced a zero for the 7. Integrating the $2x + 2$ with the antiderivative approach produced the $x^2 + 2x$ terms but not the 7. Given an integral problem as

$$y = \int (2x + 2)dx$$

the integral of $2xdx$ is x^2 and the integral of $2dx$ is $2x$ but it is impossible to determine if there is a constant in the expression for y.

The correct solution to this integral is: $y = \int (2x + 2)dx = x^2 + 2x + C$.

Quick Tip

Integrals obtained by taking the antiderivative must be written with an arbitrary constant. The constant can be determined if other details are specified in the problem. Integrals requiring a constant (of integration) are called *indefinite integrals*. There is a way around this problem but for the time being just remember to include the constant and evaluate it if possible from the information in the problem.

Example 7-3 Evaluate $y = \int(x^2 + 2x^{-2} + 3)dx$.

Solution: Follow the formula for integrating polynomials as stated earlier in this chapter or from the Mathematical Tables at the end of the book.

$$y = \frac{x^3}{3} + \frac{2x^{-1}}{-1} + 3x + C = \frac{x^3}{3} - 2x^{-1} + 3x + C$$

Example 7-4 The population of a certain region is growing with time according to $11 + 0.2\sqrt{t}$. Population is measured in thousands and time in years. The current population is 30 (thousand). What is the expression for P as a function of t?

Solution: The words "population growing with time" translated into calculus means

$$11 + 0.2\sqrt{t} = \frac{dP}{dt}$$

Writing this as an integral problem, we have $\int dP = \int(11 + 0.2t^{1/2})dt$ and

$$P = 11t + \frac{0.2t^{3/2}}{3/2} + C = 11t + \frac{0.4}{3}t^{3/2} + C = 11t + 0.13t^{3/2} + C$$

The words "current population ... 30" mean that at $t = 0, P = 30$. Put these numbers into the general expression for P to determine C. (If a variable such as P is given a value when $t = 0$, it is sometimes referred to as "the initial condition.")

$$30 = 11(0) + 0.13(0)^{3/2} + C \text{ makes } C = 30 \text{ so the specific relation is}$$

$$P = 11t + 0.13t^{3/2} + 30$$

Example 7-5 A certain car decelerates under braking at a rate of $16\,\text{ft/s}^2$. If the car is traveling at a speed of $60\,\text{ft/s}$ (approximately 40 miles per hour) when the brakes are applied, how far does it take the car to stop?

$$\longrightarrow \quad v = 60\,\text{ft}/\text{s}$$

$$\longleftarrow \quad a = -16\,\text{ft}/\text{s}^2 \qquad\qquad\qquad v = 0$$

Solution: A little review is in order. Position speed and acceleration were discussed in Chapter 3, Derivatives. You may want to review Examples 3-6 and 3-7 dealing with speed and acceleration. Stated in calculus terminology, speed, $v = ds/dt$, is change in position with time, and acceleration, $a = dv/dt$, is change in speed with time. Keep in mind that a is measured in ft/s^2, v in ft/s, and s in ft.

In this problem start with the acceleration, which is a negative number, so the first statement of the problem is $dv/dt = -16$. The integral to find v is

$$v = -\int 16dt = -16t + C_1$$

When the brakes are applied ($t = 0$), $v = 60$ ft/s so $60 = -16(0) + C_1$ and $C_1 = 60$ so

$$v = \frac{ds}{dt} = -16t + 60$$

We are looking for the distance, not the velocity, so one more integral is in order.

$$s = \int (-16t + 60)dt = -\frac{16}{2}t^2 + 60t + C_2 = -8t^2 + 60t + C_2$$

The stopping distance s is measured from where, and when, the brakes are applied so at $t = 0$, $s = 0$. This fact allows evaluation of C_2.

$$0 = -8(0)^2 + 60(0) + C_2 \quad \text{so} \quad C_2 = 0 \quad \text{and} \quad s = -8t^2 + 60t$$

To recap what we have done so far, we started with the acceleration, $a = -16$, integrated to get the speed, $v = -16t + 60$, and integrated again to get $s = -8t^2 + 60t$. All this work and we still don't have the stopping distance!

A little more logic provides the final answer. The stopping distance s could be evaluated if we knew the braking time. But the time can be determined from the speed statement. When the braking has gone on long enough, the car stops (setting $v = 0$ in $v = -16t + 60$ produces the time to stop).

$$0 = -16t + 60 \quad \text{or} \quad t = 60/16 = 3.8$$

The stopping distance, using this time, is

$$s = -8t^2 + 60t = -8(3.8)^2 + 60(3.8) = -116 + 228 = 112$$

$$0 = -6(0)^2 + 50(0) + C_2 \quad \text{so} \quad C_2 = 0$$

The problems so far have been in the form y equals the integral of some polynomial in x times dx. The next problem illustrates a type of problem where the derivative depends on both variables.

Example 7-6   The rate of change of a certain variable x with y is equal to the square root of the product of x and y. Find y as a function of x.

This is a problem where the words prescribe the mathematics. The phrase "the rate of change of . . . x with y" means derivative; the phrase "the square root of the product" is clear. Form the product and take the square root. Read the sentence carefully, several times if necessary, and write

$$\frac{dx}{dy} = \sqrt{xy}$$

With the problem written down, another difficulty appears. This is not a simple dy equals a polynomial times dx problem. Separating the variables is going to take a little work. With a little manipulation the statement can be written as

$$\frac{dx}{\sqrt{x}} = \sqrt{y}dy$$

This process is called separating the variables. While this problem is a little different from previous problems neither integral is difficult.

$$x^{-1/2}dx = y^{1/2}dy \quad \text{or} \quad \int x^{-1/2}dx = \int y^{1/2}dy$$

Perform each integration, $\dfrac{x^{1/2}}{1/2} = \dfrac{y^{3/2}}{3/2} + C$ and with a little algebra $x = \ldots$ or

$y = \ldots$ can be written as $x^{1/2} = \dfrac{y^{3/2}}{3} + C_1$ or $x = \left[\dfrac{y^{3/2}}{3} + C_1\right]^2$ or $y^{3/2} = 3x^{1/2} + C_2$

or $y = [3x^{1/2} + C_2]^{2/3}$

Notice that instead of writing $C/2$, a new constant C_1 was introduced. If at the end of the problem the constant is evaluated it does not matter whether the constant is two times the original or any other multiple or root, or whatever of the original. Also notice that C_2 is a new constant.

Example 7-7   The rate of change of y with x is equal to the product of x and y squared. Find the equation relating y and x.

Solution: Write the words in mathematics (calculus-speak).

$$\frac{dy}{dx} = xy^2$$

Separate the variables and perform the integration.

$$\frac{dy}{y^2} = x\,dx \quad \Rightarrow \quad \int y^{-2}dy = \int x\,dx \quad \Rightarrow \quad \frac{y^{-3}}{-3} = x^2 \quad \Rightarrow \quad x^2y^3 = -\frac{1}{3} + C$$

Example 7-8   Due to an unusually favorable habitat the deer population in a certain region is growing at an average rate of $0.08t + 5$ thousand per month. Find the formula for population. The present population is 200,000. If the rate continues, what will be the population in 6 months?

The rate stated in the problem is dP/dt so

$$\frac{dP}{dt} = 0.08t + 5 \quad \text{or} \quad dP = (0.08t + 5)dt$$

and integrating

$$P = \int (0.08t + 5)dt = 0.08\frac{t^2}{2} + 5t + C = 0.04t^2 + 5t + C$$

Use $t = 0$ and $P = 299$ to evaluate C (the original equation was in thousands):

$$200 = 0.04(0)^2 + 5(0) + C \quad \text{so} \quad C = 200$$

and the population formula for this region is

$$P(t) = 0.04t^2 + 5t + 200$$

In 6 months, $t = 6$, the population will be

$$P(6) = 0.04(6)^2 + 5(6) + 200 = 1.44 + 30 + 200 = 232.44 \text{ thousand}$$

Just as there was a "special" derivative associated with $\ln x$ there is a "special" integral associated with $1/x$ or x^{-1}. The integral of $1/x$ or x^{-1} is $\ln|x|$. This is the natural logarithm, or base e, logarithm and that the absolute value is required. There are no logarithms of negative numbers—try taking the ln of a negative number on your calculator. The formal definition of this integral is

$$\int \frac{1}{x}dx = \ln|x| + C$$

With this definition, do some problems.

Example 7-9 Find $\int \frac{x + 1}{x} dx.$

Solution: This is one of those innocent looking little problems that will drive you crazy if you don't know how to handle the fraction $(x + 1)/x$. The formal technique is called partial fractions but for this simple fraction all that is necessary is a little logic. Fractions often come from, or at least they can be written as, other fractions.

$$\frac{x + 1}{x} = \frac{x}{x} + \frac{1}{x} = 1 + \frac{1}{x}$$

If you can see to write the fraction this way, then the problem is not so bad. And after you have done one or two of these "fraction" problems you will remember the little "trick." Application of the "trick" is not difficult. It is knowing to use it that is important and you now know that. With this algebraic rearrangement the integral is

$$\int \frac{x + 1}{x} dx = \int \left(1 + \frac{1}{x}\right) dx = \int dx + \int \frac{1}{x} dx = x + \ln |x| + C$$

Example 7-10 Find $\int \frac{3x^3 + 2x^2 + x}{x^3} dx.$

Solution: Based on your experience from the previous problem write this fraction as three fractions:

$$\frac{3x^3 + 2x^2 + x}{x^3} = \frac{3x^3}{x^3} + \frac{2x^2}{x^3} + \frac{x}{x^3} = 3 + \frac{2}{x} + \frac{1}{x^2}$$

Now perform the three integrals to solve the problem.

$$\int \frac{3x^2 + 2x^2 + x}{x^3} = \int \left(3 + \frac{2}{x} + x^{-2}\right) dx = 3x + 2\int \frac{dx}{x} + \int x^{-2} dx$$

$$= 3x + 2 \ln|x| - x^{-1} + C$$

Partial fractions is a very powerful tool for handling the integration of polynomials over polynomials. In a polynomial where the degree of the numerator is less than the degree of the denominator and when the denominator can be factored into first degree factors, with no repeat factors, the polynomial can be reduced by a simple method of partial fractions. This is shown in the next problem.

Example 7-11 Find $\int \dfrac{1}{x^2 + x - 2} dx$.

Solution: The denominator can be factored into $(x + 2)(x - 1)$. Now set up a fraction statement where there are undetermined constants in the fractions.

$$\frac{1}{(x + 2)(x - 1)} = \frac{A}{x + 2} + \frac{B}{x - 1}$$

Multiply so as to clear the fractions.

$$1 = A(x - 1) + B(x + 2) \quad \text{so} \quad A + B = 0 \quad \text{and} \quad -A + 2B = 1$$

Add these last two equations so $3B = 1$ or $B = 1/3$ and $A = -1/3$. This produces the partial fractions and allows the integral to be written as

$$-\frac{1}{3}\int \frac{dx}{x + 2} + \frac{1}{3}\int \frac{dx}{x - 1} = \frac{1}{3}\Big[\ln |x - 1| - \ln |x + 2|\Big] + C$$

Example 7-12 The rate at which algae are growing in a certain pond is proportional to the amount of algae according to $dA/dt = 0.02A$, where A is measured in pounds and t in days. At present there is 300 lb of algae in the pond. Find the time for the amount of algae to double.

Solution: The rate statement is $dA/dt = 0.02A$. The integral of this statement is accomplished after separating the variables

$$\frac{dA}{A} = 0.02dt \quad \text{or} \quad \int \frac{dA}{A} = \int 0.02dt \quad \text{and the integration is} \quad \ln |A| = 0.02t + C$$

At $t = 0$, there is 300 lb of algae in the pond so

$$\ln 300 = 0.02(0) + C \quad \text{which makes} \quad C = \ln 300.$$

Don't worry about finding a number for $\ln 300$, $\ln 300$ is a perfectly good constant.

The statement connecting A to t is

$$\ln A - \ln 300 = 0.02t$$

The doubling time is when $A = 600$, double the original amount, so write

$$\ln 600 - \ln 300 = 0.02t \quad \text{or} \quad 6.39 - 5.70 = 0.69 = 0.02t \quad \text{and} \quad t = 34.5 \text{ days}$$

(Remember also that $\ln 600 - \ln 300 = \ln (600/300) = \ln 2 = 0.69$.)

There is one other integral formula to add to our growing collection and that is the formula for the integration of the exponential function. It is fairly simple.

$$\int e^x dx = e^x + C$$

This formula becomes particularly useful as the exponent becomes more complicated. These more difficult situations will be taken up in the next chapter.

Example 7-13   Find $y = \int e^x dx$ when $y = 3$ and $x = 0$.

Solution: The integral is $y = e^x + C$. Set $y = 3$ and $x = 0$ to find C. Remember: anything raised to the zero power is 1.

$$3 = e^0 + C \quad \text{or} \quad 3 = 1 + C \quad \text{so} \quad C = 2 \quad \text{and} \quad y = e^x + 2$$

This is a good place to stop and take another look at this process called integration. The antiderivative and formula approach work well on many problems. The next approach, the area under the curve, has some distinct advantages in certain problems. After going through the area under the curve view of integration you will be able to switch back and forth choosing which view is most convenient for a particular problem.

7-2   Area Under the Curve

Integrals are often introduced as a means of measuring the area under a curve. In certain problems the area under a curve has physical meaning and is very helpful in understanding the problem. Rather than doing a formal derivation relating the integral to the area under a curve we will show how the area is consistent with the antiderivative approach. And as usual we will do this in the context of solving problems.

Example 7-14   Find the area under the curve $y = 4$, bounded by $x = 0$ and $x = 5$.

Solution: Graph the function. It is a straight line at $y = 4$, parallel to the x-axis. To find the area, integrate $4dx$ from $x = 0$ to $x = 5$.

This area integral is written as

$$A = \int_0^5 4dx$$

The 0 and 5 mean, evaluate the integral at 5 and then subtract the value for 0. The operations are

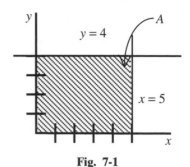

$$A = \int_0^5 4dx = 4x|_0^5 = 4(5) - 5(0) = 20$$

The shaded rectangular area shown as shaded in Fig. 7-1 has dimensions 4 by 5 and area 20, the value obtained with this integration.

Fig. 7-1

Integrals written with "limits" on the integral sign are called *definite integrals*. Since these limits clearly define the extent of the area represented, the integral does not need an arbitrary constant.

Example 7-15 Find the area under the curve $y = x$ between $x = 0$ and $x = 4$.

Solution: First, graph the curve. The area, by integration is

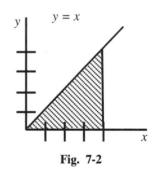

$$A = \int_0^4 xdx = \frac{x^2}{2}\bigg|_0^4 = 8 - 0 = 8$$

This curve $y = x$ forms a triangle with the x-axis and the line $x = 4$ (Fig. 7-2). The area of this triangle is one-half the base times the height $(1/2)(4)(4) = 8$, the same value as obtained through integration.

Fig. 7-2

Example 7-16 Find the area under the curve $y = x^2$ between $x = 0$ and $x = 2$.

Solution: First, graph the curve (Fig. 7-3). Using the integral approach the area is

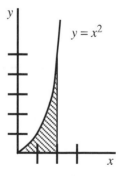

$$A = \int_0^2 x^2dx = \frac{x^3}{3}\bigg|_0^2 = \frac{2^3}{3} - 0 = \frac{8}{3} \approx 2.7$$

The area under this curve is less than the area within a triangle formed with base along the x-axis from 0 to 2, height from $y = 0$ to 4 and the slant height

Fig. 7-3

from the point (0,0) to (2,4). Such a triangle has area (1/2)2·4 = 4, and as expected is more than the area of ≈ 2.7 computed with the integral.

The curve $y = x^2$ goes through the points (1,1) and (2,4) so approximate the area under the curve with a triangle and trapezoid as shown in Fig. 7-4. The area of the triangle is (1/2)(1)(1) = 1/2. The area of the trapezoid is (1/2)(sum of opposite faces)(height) which in this case is (1/2)(1 + 4)(1) = 2.5. The sum of these areas is 3, even closer to the area of ≈ 2.7 obtained through the integral.

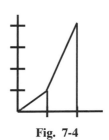

If this process were continued with narrower and narrower trapezoids the area would approach the 2.7 value obtained through the integral.

Fig. 7-4

These three problems all point toward an interpretation of the integral of a function as the area under the graph of that function over the prescribed limits.
The successive approximations of narrower and narrower trapezoids, or rectangles, leading to the area under the curve is the classic definition of the integral.

Use the same curve $y = x^2$ as an example, though any curve would work as well, and look to approximating the area not with trapezoids, but with a collection of narrow rectangles. The rectangles can be constructed in a variety of ways, inside the curve, outside the curve or using a mid value (see Fig. 7-5). It really doesn't make any difference how they are constructed because we are going to take the limit by making their width go to zero. The ones shown here are an average height. Look at the x_n'th rectangle of width Δx that has height x_n^2.

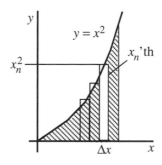

Fig. 7-5

The area under this curve can be written as a sum of similar rectangles. With this view, the area under the curve is

$$A \approx \sum_{n}(x_n^2)\Delta x$$

with the area getting closer and closer to the actual area as the width of the rectangles decreases and their number increases.

Using a limit approach, and the knowledge that the integral over a specified range in x is the area under the curve, A is the limit of the sum as Δx goes to zero.

$$A = \lim_{\Delta x \to 0}\sum_{n}(x_n^2)\Delta x = \int_0^x x^2 dx$$

The integral is viewed as the area generated by summing an infinite number of rectangles of infinitely small width.

Example 7-17   Find the area under the curve $y = x^3 - 1$ from $x = 1$ to $x = 3$.

Solution: This is a cubic. It rises steeply, and it crosses the y-axis at -1 and the x-axis at 1 and goes through the point (3,8) (see Fig. 7-6). The rectangle represents one of the rectangles that is being summed in the integration process. The shaded area is

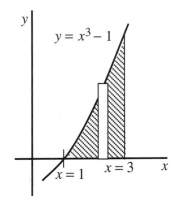

Fig. 7-6

$$A = \int_1^3 (x^3 - 1)dx = \left[\frac{x^4}{4} - x\right]_1^3$$

$$A = \left(\frac{81}{4} - \frac{12}{4}\right) - \left(\frac{1}{4} - \frac{4}{4}\right) = \frac{69}{4} + \frac{3}{4} = \frac{72}{4} = 18$$

The next several problems will explore some of the unique uses of integral calculus for finding areas.

Example 7-18   Find the area bounded by $y = 2 - (1/2)x^2$ and the x-axis.

Solution: First graph the function. If you have the least bit of problem graphing this function go to previous sections concerned with graphing parabolas. This function is a parabola. It opens down and crosses the y-axis at $y = 2$ (Fig. 7-7).

The limits on the integral have to be from where the curve crosses the x-axis on the negative side to where it crosses on the positive side. To find these points set $y = 0$ and solve for x.

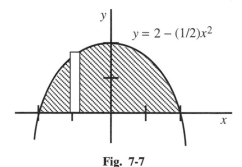

$$0 = 2 - (1/2)x^2 \quad \text{or} \quad 2 = (1/2)x^2$$
$$\text{or} \quad 4 = x^2 \text{ or } x = \pm 2$$

Fig. 7-7

The shaded area is the area desired so the integral is

$$A = \int_{-2}^{2} [2 - (1/2)x^2]dx = \left[2x - \frac{x^3}{6}\right]_{-2}^{2} = \left[2(2) - \frac{2^3}{6}\right] - \left[2(-2) - \frac{(-2)^3}{6}\right]$$

$$A = \left[4 - \frac{4}{3}\right] - \left[-4 - \frac{-8}{6}\right] = \left[\frac{12}{3} - \frac{4}{3}\right] - \left[-\frac{12}{3} + \frac{4}{3}\right] = \left[\frac{8}{3}\right] - \left[-\frac{8}{3}\right] = \frac{16}{3}$$

Second Solution: There is a little faster, a little easier, and a little less prone to error way of doing this problem. Remember the symmetry that was so helpful in graphing parabolas? Not only is there a symmetry in the graph of the curve between 0 and 2 and 0 and −2, but the area under the curve from 0 to 2 is the same as the area under the curve from 0 to −2. Therefore, the entire area between this curve and the x-axis is twice the area calculation between $x = 0$ and $x = 2$. Notice the lack of negative numbers in this solution.

$$A = 2\int_{0}^{2} [2 - (1/2)x^2]dx = 2\left[2x - \frac{x^3}{6}\right]_{0}^{2}$$

$$= 2\left\{\left[2(2) - \frac{2^3}{6}\right] - [0]\right\} = 2\left\{4 - \frac{4}{3}\right\} = \frac{16}{3}$$

In doing area problems look for symmetry that will make the problem easier and cut down on the amount of numbers you have to manipulate.

***Example* 7-19** Find the area between the coordinate axes and the curve $y = \sqrt{x} - 2$. (See Fig. 7-8.)

Solution: This has got to be an odd looking curve. Start by looking at where the curve crosses the axes. At $x = 0$, $y = -2$ and at $y = 0$, $x = 4$. One other point, $x = 1$, $y = -1$, is sufficient, along with the points where the curve crosses the axes, to sketch the curve.

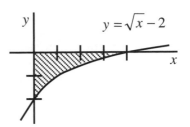

The shaded area is the only area between the curve and the axes. The area is

Fig. 7-8

$$A = \int_0^4 (x^{1/2} - 2)dx = \left[\frac{x^{3/2}}{3/2} - 2x\right]_0^4$$

$$= \left[\frac{2(4)^{3/2}}{3} - 2(4)\right] - [0]$$

$$= \frac{16}{3} - 8 = \frac{16}{3} - \frac{24}{3} = -\frac{8}{3}$$

The area is negative! Are areas below the axis negative? Just to be sure, change the limits on the integral to 4 and 5 and see if that area comes out positive, as we would expect from the graph.

$$A = \int_4^5 (x^{1/2} - 2)dx = \left[\frac{x^{3/2}}{3/2} - 2x\right]_4^5 = \left[\frac{2(5)^{3/2}}{3} - 2(5)\right] - \left[\frac{2(4)^{3/2}}{3} - 2(4)\right]$$

$$A = \left[\frac{2(5^{3/2} - 4^{3/2})}{3}\right] - 10 + 8 = \left[\frac{2(11.2 - 8)}{3}\right] - 2 = 2.1 - 2 = 0.1$$

This area comes out positive and very small, about as expected considering the curve.

The previous example illustrates an important point. Be careful when finding an area below the axis. You can end up with a negative number for the area. The following problem, Example 7-20, is a typical test problem involving positive and negative area. There is a simple way to handle this negative area situation as illustrated in this problem.

Example 7-20 Calculate the area between the curve $y = x^2 + x - 2$ and the x-axis between $x = 0$ and $x = 2$.

Solution: Do not write down the integral of $x^2 + x - 2$ with the prescribed limits and perform the integration to find the answer.

If you do, you will get the problem wrong!

Sketch the graph of this function. Factor $y = x^2 + x - 2 = (x - 1)(x + 2)$ and notice that the curve crosses the x-axis at $x = 1$ and $x = -2$. Look at the limits of the integration. At $x = 0$, $y = -2$. At $x = 2$, $y = 4$. With this information the curve can be sketched (Fig. 7-9). More detail for the sketch is not necessary.

The area between this curve and the x-axis has to be calculated in two pieces corresponding to the two areas marked A_1 and A_2.

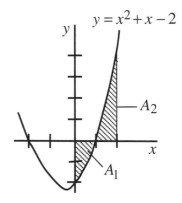

Fig. 7-9

$$A_1 = \int_0^1 [0 - (x^2 + x - 2)]dx = -\left[\frac{x^3}{3} + \frac{x^2}{2} - 2x\right]_0^1$$

$$A_1 = -\left[\frac{1^3}{3} + \frac{1^2}{2} - 2(1)\right] + [0] = -\left[\frac{2}{6} + \frac{3}{6} - \frac{12}{6}\right] = -\left[-\frac{7}{6}\right] = \frac{7}{6}$$

Notice the integrand is written as $[0-(x^2 + x - 2)]$. This statement is the "top curve," $y = 0$, minus the "bottom curve" $y = x^2 + x - 2$. Writing the integrand this way, top curve minus bottom curve, keeps the area positive. This is the preferred way of writing the problem. It will prove very helpful in more complicated problems.

Now find the second area, A_2. The integrand $x^2 + x - 2$ would be viewed as top curve minus bottom curve, which is 0.

$$A_2 = \int_1^2 (x^2 + x - 2)dx = \left[\frac{x^3}{3} + \frac{x^2}{2} - 2x\right]_1^2$$

$$A_2 = \left[\frac{2^3}{3} + \frac{2^2}{2} - 2(2)\right] - \left[\frac{1^3}{3} + \frac{1^2}{2} - 2(1)\right]$$

$$= \left[\frac{8}{3} - \frac{1}{3}\right] + \left[\frac{4}{2} - \frac{1}{2}\right] + [-4+2]$$

$$A_2 = \frac{7}{3} + \frac{3}{2} - 2 = \frac{14}{6} + \frac{9}{6} - \frac{12}{6} = \frac{11}{6}$$

The total area between the curve and the x-axis is the sum of these two areas.

$$A = A_1 + A_2 = \frac{7}{6} + \frac{11}{6} = \frac{18}{6} = 3$$

Second Standard Mistake: Don't make this mistake. If you take the integral of $x^2 + x - 2$ between the limits of 0 and 2 you will get an answer that is equal to $A_2 - A_1$. It will look great but it is wrong. Take the integral of $x^2 + x - 2$, use the limits 0 and 2 and verify that this is the difference in the areas and the incorrect answer. This is the kind of problem that math profs. use to separate the As from the Bs.

We've had As and we've had Bs. As are better.

These next few problems take you to another level. The "standard mistake" of the previous problem can be avoided by graphing. Likewise graphing is essential in these next few problems. As we mentioned in "A Special Message" to the *Utterly Confused Calculus Student* at the beginning of this book, graphing is one of the skills you need to do calculus problems. We keep emphasizing this point because we know that a primary source of confusion in integration is inability to visualize the problem, and you visualize problems by graphing the curves.

Example 7-21   Find the area in the positive x and y region between the curves $y = (0.5)x$ and $y = 4 - (0.5)x^2$.

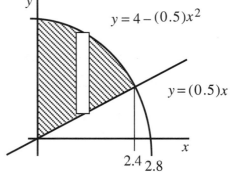

Solution: Graph the two curves in the positive x and y region (see Fig. 7-10).

The straight line is easy. The parabola is 4 at $x = 0$ and opens down. The parabola crosses the x-axis when $y = 0$ or $x^2 = 8$.

Fig. 7-10

As far as the limits of integration are concerned the important point is where the curves cross. This point is found by setting the two equations for y equal and solving for x.

There is a point along the $y = (0.5)x$ curve that satisfies $y = 4 - (0.5)x^2$. This point is where the curves cross and is found by setting $(0.5)x$ equal to $4 - (0.5)x^2$ and solving the equation

$$(0.5)x = 4 - (0.5)x^2 \quad \text{or} \quad x^2 + x - 8 = 0$$

This quadratic is solved by formula

$$x = \frac{-1 \pm \sqrt{1^2 - 4(1)(-8)}}{2(1)} = \frac{-1 \pm \sqrt{33}}{2} = 2.4$$

Only the positive root is interesting in this problem. The figure is a sketch, not a detailed drawing. The essential feature is the point where the curves cross and the visualization that the integral is over dx and between the two curves. Great detail is not necessary. A clear picture of the curves, where they cross, and the limits is sufficient information.

The integral is written as going from the top curve, or most positive part of the dx rectangle, to the bottom curve, or most negative part of the rectangle with the appropriate limits 0 and 2.4.

$$A = \int_0^{2.4} [(4 - 0.5x^2) - (0.5)x]dx = \left[4x - \frac{x^3}{6} - \frac{0.5x^2}{2}\right]_0^{2.4}$$

$$A = \left[4(2.4) - \frac{2.4^3}{6} - \frac{2.4^2}{4}\right] - [0] = [9.6 - 2.3 - 1.4] = 5.9$$

Example 7-22 Find the area bounded by the y-axis and the curves $y = 1 + \sqrt{x}$ and $y = x - 1$.

Solution: The curve $y = x - 1$ is a straight line of slope 1 that intercepts the y-axis at -1. The other curve starts at $y = 1$ and increases. To integrate in the x-direction the limits are required. In this case the upper limit in x is where the curves cross, which is obtained by setting the equations for y equal and solving

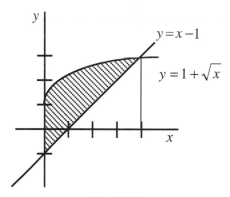

Fig. 7-11

$$1 + \sqrt{x} = x - 1 \quad \text{or} \quad \sqrt{x} = x - 2$$

and squaring

$$x = x^2 - 4x + 4 \quad \text{or} \quad x^2 - 5x + 4 = 0$$

This quadratic is factorable, $(x - 4)(x - 1) = 0$, producing values of $x = 1$ and $x = 4$.

The value $x = 1$ requires a negative square root to work in both original equations and is seen from Fig. 7-11 as incorrect. The value $x = 4$ is the correct limit value. The $x = 1$ point is a spurious one caused by squaring a square root and then factoring the resulting equation. With the limits, set up the integral from 0 to 4 of the upper curve minus the lower curve and integrate.

$$A = \int_0^4 [(1 + \sqrt{x}) - (x - 1)]dx = \int_0^4 (2 + x^{1/2} - x)dx = \left[2x + \frac{x^{3/2}}{3/2} - \frac{x^2}{2} \right]_0^4$$

$$A = \left[2(4) + \frac{4^{3/2}}{3/2} - \frac{4^2}{2} \right] = \left[8 + \frac{2}{3}(8) - 8 \right] = \frac{16}{3}$$

The next two problems are practical problems illustrating how calculus can help in forecasting revenue generation in the one instance, and yield from a mining operation in the other instance. The unique aspect of these problems is that they start not with a statement of revenue, but with a statement of revenue rate, the revenue generated per year and the yield of the mine in tons per year. Watch the way these problems are worded. Don't be fooled on a test by misreading a rate statement.

Example 7-23 A certain machine generates revenue at the rate of $R(t) = 2000 - 5t^2$, where R is in dollars per year and t is in years. As the machine ages the cost of repairs increases according to: $C(t) = 500 + 2t^2$. How long is the machine profitable and what are the total earnings to this point in time?

Solution: The two curves are both parabolas, the $R(t)$ curve opening down and the $C(t)$ curve opening up. The curves are sketched here (Fig. 7-12). Detail is not necessary.

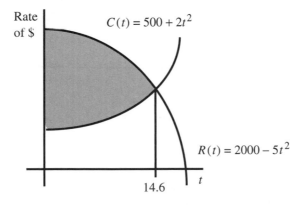

Fig. 7-12

When the revenue generated per year equals the cost of repairs per year the machine stops being profitable.

Mathematically, this situation occurs when the curves cross. The time when they cross is found by setting the equations equal and solving for the time.

$$2000 - 5t^2 = 500 + 2t^2 \quad \text{or} \quad 1500 = 7t^2$$

$$\frac{1500}{7} = t^2 \quad \text{or} \quad t = \sqrt{\frac{1500}{7}} = 14.6 \text{ years}$$

The total earnings up to 14.6 years is the (revenue generated) area under the $R(t)$ curve minus the (cost) area under the $C(t)$ curve. This is an integral. Look at the units. The rate of return in dollars per year times the time is the total number of dollars.

$$E = \int_0^{14.6} [(2000 - 5t^2) - (500 + 2t^2)]dt = \int_0^{14.6} [1500 - 7t_2]dt$$

$$E = \left[1500t - \frac{7t^3}{3} \right]_0^{14.6} = 1500(14.6) - \frac{7(14.6)^3}{3} = 21900 - 7262 = 14{,}638$$

The total earnings until the machine becomes unprofitable, that is, costs more to operate each year than it returns in revenue, is $14,638.

Example 7-24 In a mine the yield per unit cost for a particular ore is declining according to $Y = 8 - 0.4t$, where the yield is in millions of tons per year and t is in years. Find the time for the mine to produce 60 million tons of ore.

Solution: Be careful with rate statements like this one. The yield equation is in millions of tons per year, not millions of tons total. Since the yield is in millions of tons per year, the time for 60 million tons has to come from an integration over time. Integration is required rather than multiplication because the rate per year is changing. The total yield then is

$$T = \int_0^t (8 - 0.4t)dt = \left[8t - \frac{0.4t^2}{2} \right]_0^t = 8t - 0.2t^2$$

Notice that in this problem the limits are 0 and t because we are looking for the time to produce a total of 60 (million tons). Therefore, set $T = 60$ in the equation

$$T = 8t - 0.2t^2$$

generated by the integral and solve for the time.

$$60 = 8t - 0.2t^2 \quad \text{or} \quad 0.2t^2 - 8t + 60 = 0 \quad \text{or} \quad t^2 - 40t + 300 = 0$$

This quadratic is factorable to $(t - 10)(t - 30) = 0$ producing time values of 10 and 30.

Go back to the original statement for the yield $Y = 8 - 0.4t$ and note that at $t = 10$ the yield is $Y(10) = 8 - 0.4(10) = 4$ and at $t = 30$ the yield is $Y(30) = 8 - 0.4(30) = 8 - 12 = -4$.

The 10 year figure is the realistic one. Who would work the mine until the yield reached zero and then continue, putting ore back, until the 60 million total was achieved?

Further Insight Solution: If the yield is $Y = 8 - 0.4t$ then in 10 years the yield goes from 8 (starting at zero time) to $8 - 4 = 4$ in a linear fashion so the average yield over the 10 years is 6. This 6 million tons per year average times the 10 years produces the 60 million tons.

The 30 year figure is also true. If the yield goes according to $Y = 8 - 0.4t$ for 30 years then the yield goes from 8 at time zero to -4 at the end of 30 years and the average is 2 million tons per year for 30 years for the 60 million ton total.

No one would actually do this because when the yield went to zero you would have to start putting ore back into the mine to achieve your 60 million tons total! You would also expect the yield equation to not accurately represent the mine production after the production rate had gone to zero.

Sometimes, in problems involving quadratics, solutions are generated that are mathematically correct but unrealistic. It is good practice to always look at the answer and ask if it is reasonable.

Example 7-25   A demographic study indicates that the population of a certain town is growing at the rate of $4 + 2x^{0.8}$ people per month, when x is measured in months. What will be the increase in population between the 10th and 12th months?

Solution: This is an integral problem. The growth function has to be integrated and evaluated at the 10th and 12th months. Write the integral as the number (10 − 12) and use the growth function integrated over time.

$$N(10 - 12) = \int_{10}^{12} (4 + 2x^{0.8})dx = \left[4x + \frac{2x^{1.8}}{1.8} \right]_{10}^{12}$$

$$N(10 - 12) = \left[4(12) + \frac{2(12)^{1.8}}{1.8} \right] - \left[4(10) + \frac{2(10)^{1.8}}{1.8} \right]$$

Before going any further review how to take a fractional power with your calculator. To find $(12)^{1.8}$ enter 12 on your calculator, then find a key that raises "things" to a power (this key will look like y^x or x^y) and press it. Your calculator will probably blink and continue to display the 12. That's OK. Don't worry about the 12, enter 1.8, and press the equal sign. The calculator should take a short time to display 88.

$$N(10 - 12) = \left[48 + \frac{2(88)}{1.8} \right] - \left[40 + \frac{2(63)}{1.8} \right] = [48 + 98] - [40 + 70] = 36$$

A total of 36 people will enter the town in the 10th to 12th month interval.

Example 7-26 A rare stamp is, and has been, appreciating at the rate of $5 + 0.5t$ in thousands of dollars per year when t is measured in years. If this stamp is purchased for $5000 for a newborn child and allowed to appreciate, what will be the value of the stamp on the child's 18th birthday?

Solution: This is a rate problem and an integral is required. The stamp is purchased (at $t = 0$) for $5000. Integrate the rate over time with the limits of 0 and 18 to find the value after 18 years.

$$V = \int_{0}^{18} (5 + 0.5t)dt = \left[5t + \frac{0.5t^2}{2} \right]_{0}^{18} = \left[5(18) + \frac{0.5(18)^2}{2} \right] = [90 + 81] = 171$$

In 18 years the stamp will be worth $171,000.

7-3 Average Value of a Function

Integral calculus can be used to determine the average values of functions. The average value of some quantity that may be varying in a very complicated way can be a valuable piece of information. The average value of a function is the area under the curve of that function over a certain range divided by that range. The area under the curve is viewed as the area of a rectangle with one dimension equal to the range of the integral and the other dimension, the height equal to the average height to produce the area under the curve.

The formal definition is

$$\text{Average value} = \frac{1}{b-a} \int_a^b f(x)dx$$

The several problems in this section show how to find the average value of several different functions and illustrate applications of the technique.

Example 7-27 On an employee stock purchase plan one share of stock is purchased each month for 10 months. The share prices start at $10 at the end of the first month and decrease by $1 per month thereafter for the duration of the offer. This is an incentive (to stay with the company) plan and it does not reflect the actual stock price.

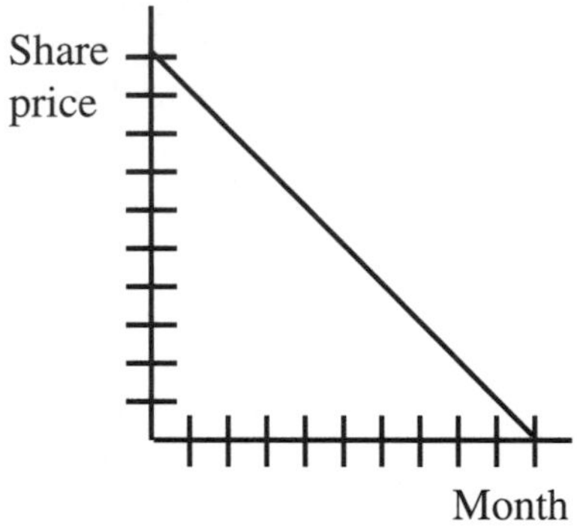

Fig. 7-13

Solution: You don't need calculus to do this problem. Graph the stock purchase price as in Fig. 7-13. Look at the graph and conclude that the average purchase price is $5 over the 10 month interval for a total cost of $50 for the 10 shares.

Think Calculus Solution: The area enclosed by the triangle in Fig. 7-13 represents the total cost for the 10 shares of stock, $50. This area is also (1/2) base × height = (1/2)(10)(10) = 50. This area could be represented by a rectangle of the same base and height 5. The height of 5 is an average height of the triangle. In mathematical language the height of the rectangle would be

$$\text{Height} = \frac{\text{area of rectangle}}{\text{base of rectangle}}$$

Example 7-28 In another stock purchase plan one share of stock is offered each month starting at $40. The history of the stock indicates the price will follow $C = 40 + 0.8t$ for the next year where t is in months. If 12 shares are purchased according to this plan, what will be the average price of the stock?

Solution: Graph the price of the stock as shown in Fig 7-14. The area under the curve is the total cost for the 12 shares. This (total cost) area divided by 12, the base of the rectangle with area equivalent to this total cost, gives the average price of the stock.

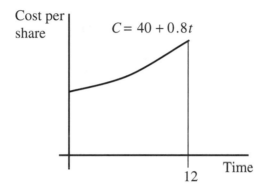

Fig. 7-14

Following the form of integral stated earlier we calculate

$$C_{avg} = \frac{1}{12 - 0} \int_0^{12} (40 + 0.8t)\,dt$$

$$C_{avg} = \frac{1}{12}\left[40t + \frac{0.8t^2}{2} \right]_0^{12}$$

$$C_{avg} = \left[40(12) + \frac{(0.8)(12)^2}{2} \right]$$

$$C_{avg} = \frac{1}{12}[480 + 57.6] = 44.8$$

The average price of the stock will be $44.80.

The average-value-of-a-function problems so far have had pretty tame-looking functions. This next problem will illustrate how to apply the average value of a function to some more complex functions.

Example 7-29 Find the average value of the function $y = x^3 - 2x^2 + 3$ from $x = -1$ to $x = 3$.

Solution: It is important to graph this function, or at least put in some values so we know whether the function is positive or negative over the region. In some problems it may be perfectly acceptable for the values of the function to be negative, while in other functions we may be confined to averaging only positive values. The dominant term is the cubic so for large x the curve has the cubic shape (see Chapter 1, Mathematical Background). A third-degree equation has at most two points where the slope is zero (see Chapter 4, Graphing).

Since only a rough sketch is necessary perhaps it will prove sufficient to just find a few points and place them on the graph.

$y(0) = 0^3 - 2(0)^2 + 3 = 3$

$y(1) = 1^3 - 2(1)^2 + 3 = 2$

$y(2) = 2^3 - 2(2)^2 + 3 = 3$

$y(3) = 3^3 - 2(3)^2 + 3 = 12$

$y(-1) = (-1)^3 - 2(-1)^2 + 3 = 0$

This function is positive over the range where it is to be averaged. Don't be fooled by an exam question that asks you to average all the positive values for a function over a certain range and then gives you a function that is negative over part of the range.

The shaded area in Fig. 7-15, the prescribed region in x, is all above the axis so the integral for the average value of the function can be written knowing that there will not be a negative area.

$$y_{avg} = \frac{1}{3 - (-1)} \int_{-1}^{3} (x^3 - 2x^2 + 3)dx = \frac{1}{4}\left[\frac{x^4}{4} - \frac{2x^3}{3} + 3x \right]_{-1}^{3}$$

$$y_{avg} = \frac{1}{4}\left\{ \left[\frac{3^4}{4} - \frac{2(3)^3}{3} + 3(4) \right] - \left[\frac{(-1)^4}{4} - \frac{2(-1)^3}{3} + 3(-1) \right] \right\}$$

$$y_{avg} = \frac{1}{4}\left\{ \left[\frac{81}{4} - \frac{54}{3} + 12 \right] - \left[\frac{1}{4} + \frac{2}{3} - 3 \right] \right\} = \frac{1}{4}\left\{ \frac{80}{4} - \frac{56}{3} + 15 \right\}$$

$$y_{avg} = \frac{1}{4}\left\{ 35 - \frac{56}{3} \right\} = \frac{1}{4}\left\{ \frac{105}{3} - \frac{56}{3} \right\} = \frac{1}{4}\left\{ \frac{49}{3} \right\} = \frac{49}{12} = 4.1$$

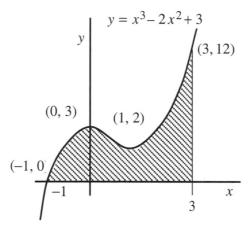

Fig. 7-15

Look at the graph and ask if this is reasonable. This average value means that the rectangle equivalent to the area under this curve would have base 4 and height 4.1, which looks very reasonable. The mistake you are looking for here is a sign mistake amongst the fractions or forgetting the (1/4) outside the whole integral.

7-4 Area between Curves Using *dy*

All of the area under the curve and average value of a function problems encountered so far have been ones where the integration was carried out in the *x*-direction. There are problems where this is inconvenient or even impossible, and it is necessary to integrate in the *dy* direction. This takes a little reorientation from the usual. In addition, the integrals are often more difficult. These problems tend to separate the *A*s from the *B*s. Follow through the several examples and learn how to find areas using integration in the *y*-direction as well as the *x*-direction.

Example 7-30   Find the area bounded by the curves $x = y^2$ and $x = 4$.

Solution: The curve $x = y^2$ is a parabola, but it is an unusual one in that it is written $x = y^2$, rather than the more familiar $y = x^2$. This means that the parabola is symmetric about the *x*-axis rather than the *y*-axis. The two curves $x = y^2$ and $x = 4$ are graphed in Fig. 7-16.

Imagine placing a representative rectangle of width *dx* on this graph. There is a problem almost immediately. The rectangle doesn't go from one curve to another. It begins and ends on the same curve!

You could solve $x = y^2$ for y to get $y = \sqrt{x}$ and then use a symmetry argument and say that the desired area is twice the area between the curve $y = \sqrt{x}$, $y = 0$, and $x = 3$. This would work for this particular problem but with only a slight modification to the parabola (add a constant, for instance) the solution for y becomes most complicated.

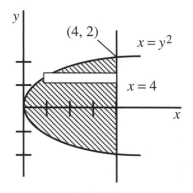

Fig. 7-16

Using a rectangle of width dy is much easier. Draw a rectangle as shown in Fig. 7-16 and integrate over dy. The most convenient limits of the integral are $y = 0$ and $y = 2$, the top half of the desired area. The shaded area is then twice this integral.

$$A = 2 \int_0^2 y^2 dy = \left[\frac{2y^3}{3} \right]_0^2 = \frac{16}{3}$$

Example 7-31   Find the area bounded by $x = y^2$, $y = -x + 3$, and $y = 0$.

Solution: To do this problem in dx would require two separate integrals, one from 0 out to the value of x for the intersection of $y = -x + 3$ and $x = y^2$, and another from this point out to $x = 3$.

It is easier to integrate in the y-direction. This integral is no longer the "top curve" minus the "bottom curve" but the "most positive in x curve" minus the "least positive in x curve."

The "most positive in x curve" is $y = -x + 3$, which has to be rewritten as $x = 3 - y$. (To integrate in the y-direction, the equations have to be in terms of ys.) The "least positive in x curve" is $x = y^2$. Figure 7-17 shows the curves and the rectangle.

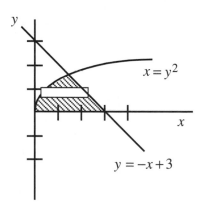

Fig. 7-17

The limits for y are zero and the value of y where the line $x = 3 - y$ and the parabola $x = y^2$ intersect. This intersection point is obtained by setting these two equations equal and solving for y.

$$3 - y = y^2 \quad \text{or} \quad y^2 + y - 3 = 0$$

This quadratic has to be solved by formula:

$$y = \frac{-1 \pm \sqrt{1 - 4(1)(-3)}}{2(1)} = \frac{-1 \pm \sqrt{13}}{2} = 1.3, -2.3$$

The positive root, 1.3, is the one for this point.

The integral for the shaded area is

$$A = \int_0^{1.3} [(3 - y) - y^2]dy$$

$$A = \left[3y - \frac{y^2}{2} - \frac{y^3}{3} \right]_0^{1.3}$$

$$A = \left\{ \left[3(1.3) - \frac{(1.3)^2}{2} - \frac{(1.3)^3}{3} \right] - [0] \right\}$$

$$A = 3.9 - 0.84 - 0.73 = 2.3$$

Example 7-32 Find the area between $x = y^2$ and $y = x - 2$.

Solution: First graph the parabola $x = y^2$ and the line $y = x - 2$ as shown in Fig. 7-18. This is one of the more difficult problems in area between two curves because of the little piece of the area near the apex of the curve. An integration in x is incorrect because in this piece of area near the apex of the curve you would be integrating between the same curve. This integration must be done in the y-direction if it is to be performed with one integral.

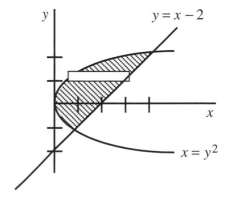

Fig. 7-18

Rewrite the line as $x = y + 2$ and set this equal to $x = y^2$ to find the values of y where the curves intersect.

$y^2 = y + 2$ or $y^2 - y - 2 = 0$ or $(y - 2)(y + 1) = 0$ producing values of $y = 2$ and $y = -1$.

The points where the curves intersect are obtained from either equation: $(2,4)$ and $(1,-1)$.

The integral for the desired area is

$$A = \int_{-1}^{2} [(y + 2) - y^2]dy$$

$$A = \left[\frac{y^2}{2} + 2y - \frac{y^3}{3} \right]_{-1}^{2}$$

$$A = \left\{ \left[\frac{2^2}{2} + 2(2) - \frac{2^3}{3} \right] - \left[\frac{(-1)^2}{2} + 2(-1) - \frac{(-1)^3}{3} \right] \right\}$$

$$A = \left\{ \left[2 + 4 - \frac{8}{3} \right] - \left[\frac{1}{2} - 2 + \frac{1}{3} \right] \right\}$$

$$A = 6 - \frac{8}{3} - \frac{1}{2} + 2 - \frac{1}{3} = 8 - \frac{9}{3} - \frac{1}{2} = 5 - \frac{1}{2} = \frac{9}{2}$$

Example 7-33 Find the area between $y = \sin x$ and the x-axis from $x = 0$ to $x = \pi$.

Solution: Graph $y = \sin x$ from $x = 0$ to $x = \pi$ as shown in Fig. 7-19. Here is another instance where symmetry can be used in calculating the area.

The area between $x = 0$ and $x = \pi/2$ is twice the area between $x = 0$ and $x = \pi$. Writing the area in the form of an integral,

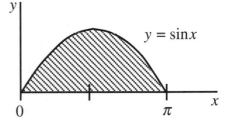

Fig. 7-19

$$A = 2 \int_{0}^{\pi/2} \sin x dx = -2\cos x \Big|_{0}^{\pi/2} = -2[0-1] = 2$$

The easiest way to verify this integral is to refer to the Mathematical Tables. If you have any trouble recalling the shape of the cosine curve, check Chapter 1, Mathematical Background.

It's a Wrap

✔ The antiderivative: an educated guessing game

✔ Area under the curve in both dx and dy directions

✔ Use integrals to find average values of functions

Test Yourself

PROBLEMS

1. Find $\int y^5 dy$.

2. Find $\int (x^2 + x^{-3} + 3)dx$.

3. The racoon population in a park is growing according to $dR/dt = 2 + 0.2t^{1/4}$, where R is the number of racoons in hundreds and t is the time in years. If the current population is 260 racoons, what will be the population at the end of 4 years?

4. A coconut is thrown down from a cliff with an initial velocity (down) of 5.0 m/s. How far does the coconut travel in 3.0 s. Take the acceleration of gravity as 9.8 m/s^2.

5. Solve $dy/dx = x^2y^2$.

6. Solve for x: $\int \dfrac{1}{x + 1}dx$.

7. Solve $\int \dfrac{x^2 + 2x + 1}{x^2}dx$.

8. Find the area between the x-axis and $y = -x - 2$ from $x = -3$ to $x = 2$.

9. Find the area between $y = 5$ and $y = x^2$.

10. Find the average value of the function $f(x) = \sqrt{x + 1}$ from $x = 0$ to $x = 3$.

11. Find the area between $x = 6 - y^2$ and $x = y$.

ANSWERS

1. Follow the pattern and write $\dfrac{y^6}{6}$.

2. $\dfrac{x^3}{3} - \dfrac{x^{-2}}{2} + 3x + C$. Check by differentiating: $\frac{1}{3}3x^2dx - \frac{1}{2}(-2)x^{-3}dx + 3dx$.

3. $R(t) = 2t + 0.2\dfrac{t^{5/4}}{5/4} + C$. At $t = 0$, $R(0) = 260 = C$ so $R(t) = 2t + 0.16t^{5/4} + 260$ and at $t = 4$, $R(4) = 2(4) + 0.16(4)^{5/4} + 260 = 269$.

4. The velocity, acceleration, and distance traveled are all in the same direction, down, so take the down direction as positive. Start with acceleration, which in calculus-speak is $d(\text{velocity})/d(\text{time})$.

$$a = 9.8 \,\frac{\text{m}}{\text{s}^2} \downarrow \qquad v_0 = 5.0 \,\frac{\text{m}}{\text{s}} \downarrow$$

$$s$$

$$\frac{dv}{dt} = 9.8 \,\frac{\text{m}}{\text{s}^2} \quad \int dv = \int 9.8 \,\frac{\text{m}}{\text{s}^2} dt$$

$$v = 9.8 \,\frac{\text{m}}{\text{s}^2} t + C_1$$

At $t = 0$, $v = 5.0$ m/s $= C_1$ so

$$v = 9.8 \,\frac{\text{m}}{\text{s}^2} t + 5.0 \,\frac{\text{m}}{\text{s}}$$

Again in calculus-speak, v is ds/dt

$$ds = 9.8 \,\frac{\text{m}}{\text{s}^2} t \, dt + 5.0 \,\frac{\text{m}}{\text{s}} dt \qquad s = 9.8 \,\frac{\text{m} t^2}{\text{s}^2 \, 2} + 5.0 \,\frac{\text{m}}{\text{s}} t + C_2$$

The distance is measured positive down from the top of the cliff so at $t = 0$, $s = 0 = C_2$.

$$s(3) = 15 \,\frac{\text{m}}{\text{s}} + 44 \,\frac{\text{m}}{\text{s}} = 59 \,\frac{\text{m}}{\text{s}}$$

5. $y^{-2} dy = x^2 dx \qquad \dfrac{y^{-1}}{-1} = \dfrac{x^3}{3} + C_1 \quad -3 = y(x^3 + 3C_1)$

6. Let $u = x + 1$ and $du = dx$ so $\displaystyle\int \frac{du}{u} = \ln|u|$ so $\displaystyle\int \frac{1}{x+1} = \ln|x+1|$

7. Long divide for $\displaystyle\int \left(1 + \frac{2}{x} + \frac{1}{x^2}\right) dx = x + 2\ln|x| - x^{-1} + C$

8. Be careful of this question! If you perform the integral from -3 to 2 you will get it wrong. Do two integrals.

$$A_1 = \int_{-3}^{-2} (-x - 2)dx = \left. -\frac{x^2}{2} - 2x \right|_{-3}^{-2}$$

$$A_1 = \left[-\frac{4}{2} + 4 \right] - \left[-\frac{9}{2} + 6 \right] = \frac{1}{2}$$

$$A_2 = \int_{-2}^{3} [(0) - (-x - 2)]dx = \int_{-2}^{3} (x + 2)dx$$

$$A_2 = \left. \left[\frac{x^2}{2} + 2x \right] \right|_{-2}^{3} = \left[\frac{3^2}{2} + 2(3) \right] - \left[\frac{(-2)^2}{2} + 2(-2) \right] = \frac{25}{2}$$

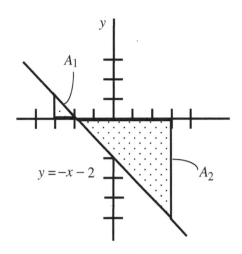

Total area is the sum of these two areas, or 13.

9. Use the symmetry and remember "upper curve − lower curve."

$$\frac{1}{2}A = \int_0^{\sqrt{5}} (5 - x^2)dx = 5x - \frac{x^3}{3}\Big|_0^{\sqrt{5}}$$

$$\left[5\sqrt{5} - \frac{5\sqrt{5}}{3}\right] - [0] = \frac{10\sqrt{5}}{3}$$

Total area is $\dfrac{20\sqrt{5}}{3}$

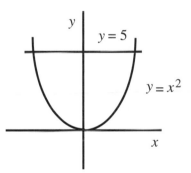

10. $f(0 - 3)_{avg} = \dfrac{1}{3 - 0}\displaystyle\int_0^3 \sqrt{x + 1}dx$

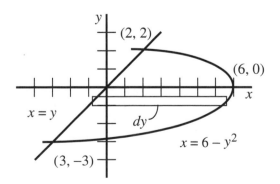

Let $u = x + 1$ so $du = dx$ and the integral is $\int u^{1/2} du = \dfrac{u^{3/2}}{3/2}$.

$$f(0 - 3)_{avg} = \frac{12}{33}(x + 1)^{3/2} \Big|_0^3 = \frac{16}{9} - \frac{2}{9} = \frac{14}{9}$$

11. This is a problem where graphing really helps. Look at the graph and realize that the integral has to be over dy. From the graph and a little algebra find where the curves intersect.

$y = 6 - y^2 \Rightarrow y^2 + y - 6 = (y + 3)(y - 2)$

At $y = 2, x = 2$

At $y = -3, x = -3$

Remember that the integral is from the most positive curve to the less positive curve.

$$A = \int_{-3}^{2} [(6 - y^2) - y]dy = 6y - \frac{y^2}{2} - \frac{y^3}{3}\Big|_{-3}^2 = \frac{125}{6}$$

CHAPTER 8

TRIGONOMETRIC FUNCTIONS

Do I Need to Read This Chapter?

You should read this chapter if you need to review or you need to learn about

➡ Right angle trigonometry

➡ Non-right angle triangles

➡ Radians and small angles

➡ Sine and cosine laws

➡ Trigonometric functions

➡ Differentiating trigonometric functions

➡ Integrating trigonometric functions

The review of the essentials of trigonometry in Chapter 1, Mathematical Background, is a review of the bare necessities for getting started in calculus. Now that you understand differentials and integrals, it is time to move on to a more complete understanding of trigonometry. This chapter covers from right angle trigonometry to the differentiation and integration of trigonometric functions. If you want a comprehensive review of trigonometry that will help you in your study of calculus this is the chapter for you. Formulas for the area and volume of geometric figures encountered in this chapter are in the Mathematical Tables at the end of the book. We begin the study of trigonometry at the very basis of trigonometry, the right triangle.

8-1   Right Angle Trigonometry

The basic right triangle is shown in Fig. 8-1. An angle and the three sides are labeled as shown. The side "opposite" is opposite the angle, whichever one it may be, and the "adjacent" is the side adjacent to the angle. The hypotenuse is always the side opposite the right angle. The little square placed in the corner indicates a right angle and the other angle is designated with a θ.

$$\sin\theta = \frac{\text{Opposite}}{\text{Hypotenuse}} = \frac{b}{c}$$

$$\cos\theta = \frac{\text{Adjacent}}{\text{Hypotenuse}} = \frac{a}{c}$$

$$\tan\theta = \frac{\text{Opposite}}{\text{Adjacent}} = \frac{b}{a}$$

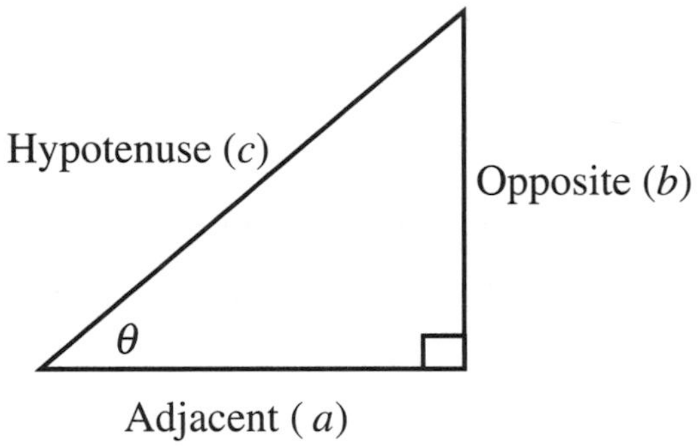

Fig. 8-1

The ratio of the sides and either side, the opposite or adjacent, to the hypotenuse is unique for each angle. These three ratio combinations are called the *sine*, *cosine*, and *tangent*. The inverses of these ratios are almost totally uninteresting.

The angles are measured in degrees with 360 degrees the total (complete circle) angle. A right angle is 90 degrees, written 90°. Each degree is further subdivided into 60 minutes, and each minute into 60 seconds. Your hand calculator probably works in degrees and decimal parts of degrees unless you have done something to make it read minutes and seconds. The minutes and seconds feature may not be available on your calculator. Most calculations are carried out to the nearest degree or nearest tenth of a degree.

The three basic ratios, the sine (sin), cosine (cos), and tangent (tan), are defined in Fig 8-1. The ability to calculate this ratio information is stored in your hand calculator. If it is not stored in your present calculator, get a better calculator. This information is so important and the calculator so inexpensive you should obtain one. If you are not familiar with how to work the calculator, practice taking a few sines, cosines, and tangents.

$$[\sin 30° = 0.50, \ \cos 75° = 0.2697, \ \tan 45° = 1.00]$$

If you did not get these numbers when you punched in $\sin 30°$ your calculator may have been in the wrong mode. Your calculator will take the sine in three different modes, degrees, rads (short for radians), and grads. Here's a simple rule. Never use grads, rarely use rads, and always check you calculator for mode. Being in the wrong mode is too embarrassing a mistake to make on a test. Actually you will, or may, use rads occasionally, but not in the context of right angle trigonometry problems. Rads will be taken up later. For now, stick to degrees.

A couple of simple problems will illustrate the use of these angle ratios in right angle trig.

Example 8-1   Fifty feet out from the base of a tree the angle measured to the top of the tree is 35°. How tall is the tree?

Solution: Figure 8-2 shows the tree, distance along the ground, the adjacent 50 ft side, and the 35° angle. The tangent function relates the two sides to the angle.

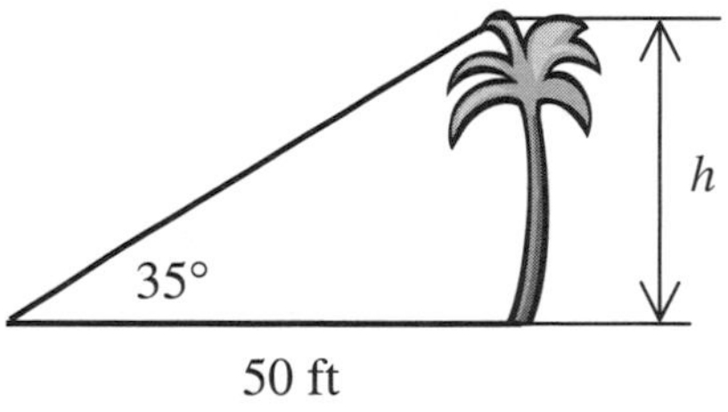

$$\tan 35° = \frac{\text{opposite}}{\text{adjacent}} = \frac{h}{50 \text{ ft}}$$

50 ft

Fig. 8-2

Solve this statement for h the same as with any algebra statement.

$$h = (50 \text{ ft}) \tan 35° = (50 \text{ ft})(0.70) = 35 \text{ ft}$$

The height of the tree is 35 ft.

Example 8-2   A certain right triangle has sides 5 and 7. Find the angles and the other side.

Solution: Sketch a right triangle and label the sides as shown in Fig. 8-3.

Start by calculating the angle θ.

$$\tan \theta = \frac{5}{7}$$

This presents a new manipulative problem in that we seek the angle with tangent ratio 5/7. On your hand calculator, enter 5 divided by 7 to display the decimal 0.71.

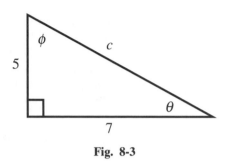

Fig. 8-3

Now perform the inverse tan function. This is usually a key labeled "inv" or "arc" or " $\tan^{-1}$," or sometimes the operation requires two keys "arc" and "tan," or "inv" and "tan." The $\tan^{-1}$ is the more popular. Pressing the appropriate key or series of keys should produce an angle of 35.5°. Remember to keep your calculator in degree mode.

The mathematical operation performed by these sequence of keystrokes is the inverse of $\tan \theta$. Take the $\tan^{-1}$ of each side of the $\tan \theta = 5/7$ equation.

$$\tan^{-1}(\tan \theta) = \tan^{-1}(5/7) \quad \text{or} \quad \theta = \tan^{-1}(0.71) = 35.5°$$

Rather than say "tangent to the minus 1" the words "arc" or "inverse" are used. The equation $\theta = \tan^{-1}(5/7)$ would be said, "theta is the arctangent of five over seven" or, "theta is the inverse tangent of five over seven."

Now calculate the angle ϕ. All the angles of the triangle have to add to 180° so 180° − 90° − 35.5° = 54.5°. The angle ϕ is 54.5°.

Find the hypotenuse using the cosine function.

$$\cos 35.5° = \frac{7}{c} \quad \text{or} \quad c = \frac{7}{\cos 35.5°} = 8.6$$

The hypotenuse is 8.6.

Another useful property of right triangles is the Pythagorean theorem. In words, the Pythagorean theorem is: in a right triangle the sum of each side (individually) squared equals the hypotenuse squared. Referring to Fig. 8-1, the theorem is written symbolically as $a^2 + b^2 = c^2$.

Example 8-3 Find all the sides and angles in a right triangle with side 4 and opposite angle 28°.

Solution: Sketch the triangle (see Fig. 8-4).

Find the hypotenuse using the sine function.

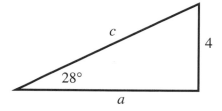

Fig. 8-4

$$\sin 28° = \frac{4}{c} \quad \text{or}$$

$$c = \frac{4}{\sin 28°} = \frac{4}{0.47} = 8.5$$

Now use the Pythagorean theorem to find the adjacent side.

$$c^2 = 4^2 + a^2 \quad \text{or} \quad a^2 = c^2 - 4^2 = 8.5^2 - 4^2 = 56.6 \quad \text{or} \quad a = 7.5$$

The other angle is $90° - 28° = 62°$.

Example 8-4 A force of 70 lb is pulling on a box sliding along a floor. The rope exerting the force is at an angle of 20° from the floor. What forces acting parallel to the floor and perpendicular to the floor would produce this force?

Solution: Finding the components of a force or speed is common in many problems. The force is viewed as having components along the floor, because that is the direction of motion of the box, and perpendicular to the floor as shown in Fig. 8-5.

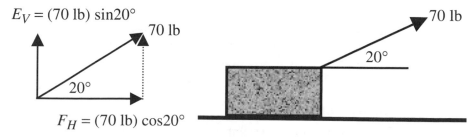

Fig. 8-5

The horizontal component of the force is $F_H = (70\,\text{lb}) \cos 20° = 66$ lb.

Write $\cos 20° = \dfrac{F_H}{70\,\text{lb}}$ and solve for F_H.

The vertical component of the force is $F_V = (70\,\text{lb}) \sin 20° = 24$ lb.

Notice that the horizontal force and the vertical force do not add up to the 70 lb. The reason for this is that they are not in the same direction. Certain measured quantities have this directional property. To describe motion or force it is necessary to add a direction. If you move 3 ft and then 4 ft you will be at very different positions relative to your starting point if you make both moves either in a straight line, at right angles to one other or first forward and then backward. Depending on the angle between the subsequent moves you will be anywhere from 1 to 7 ft from your starting point.

To describe temperature no such direction is required. Temperature is just a number, while motion requires a number plus a direction for complete description.

The components of the force are the sides of a right triangle and as such their squares should add up to the square of the hypotenuse (Pythagorean theorem).

$$a^2 + b^2 = c^2 \quad \text{or} \quad 66^2 + 24^2 = 70^2$$

8-2   Special Triangles

There are certain triangles that occur often enough to have their own names. When someone describes a problem using the phrase "similar" triangles, for example, it is important to know what that means. These definitions of triangles are not difficult, though they are sometimes difficult to remember on tests. The features and some typical uses of these triangles are shown below.

Pythagorean Triangles

Certain integral-number-sided right triangles satisfy the Pythagorean theorem. These triangles occur often enough so you should at least be aware of them. The simplest is the 3, 4, 5 triangle: $3^2 + 4^2 = 5^2$. The double of this one also works: $6^2 + 8^2 = 10^2$.

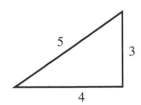

Congruent Triangles

Two triangles are congruent if they are exactly the same, sides the same, and angles the same.

Equilateral Triangles

Equilateral triangles have all their sides equal and all their angles equal. Since all three equal angles must add to 180°, the angles in an equilateral triangle are each 60°.

Isosceles Triangles

An isosceles triangle has two sides equal and the two angles opposite the equal sides also equal.

Similar Triangles

Similar triangles have the same angles. A triangle similar to another is either larger or smaller than the other. The angles are the same and the sides are in proportion. The proportion is illustrated in Fig. 8-6.

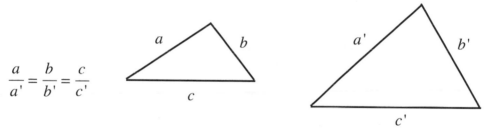

$$\frac{a}{a'} = \frac{b}{b'} = \frac{c}{c'}$$

Fig. 8-6

Example 8-5 Similar triangles often occur one inscribed inside another. For the situation shown in Fig. 8-7, find the height of the "inner" triangle.

Solution: These are similar triangles. Their angles are the same and the ratios of their sides are equal. In the larger triangle the side ratio is 10/26. In the smaller triangle this ratio is $h/20$. Set these ratios equal and solve for h.

$$\frac{h}{20} = \frac{10}{26} \quad \text{or} \quad h = \frac{(20)(10)}{26} = 7.7$$

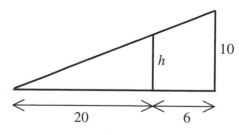

Fig. 8-7

8-3   Radians and Small Angles

Right angle trigonometry is closely related to the circle. Figure 8-8 shows a circle on a right angle coordinate system with a radius and the projection of that radius on both the *x*- and *y*-axes. The radius is 1. If the angle is measured counterclockwise from the *x*-axis (mathematicians always measure angles counterclockwise), then the sine and cosine are defined as:

$$\sin\theta = b/1 \quad \text{or} \quad b = \sin\theta$$

$$\cos\theta = a/1 \quad \text{or} \quad a = \cos\theta$$

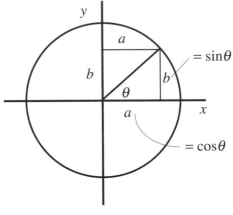

If you knew nothing about radians and degrees and were confronted with the problem of deciding how to measure an angle θ for a circle of radius 1, you probably would take the ratio of the arc length to the radius, and if the radius were 1 then the angle would be measured by the arc length. Figure 8-9 shows a radius, the angle, and the arc length. For a circle of radius 1, the circumference is 2π so a complete angle, all the way around, in this rather logical system would have an arc length of 2π. One-quarter of the way around would be a right angle and have an arc length of $\pi/2$, and so on. This arc length to radius ratio produces a pure number (no units) and defines what is known as radian measure. The relation between degrees and radians is $360° = 2\pi$ radians.

Fig. 8-8

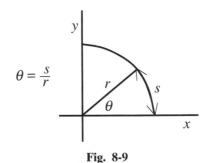

$$\theta = \frac{s}{r}$$

Fig. 8-9

A radian, because of its definition, is dimensionless so the use of the word radian or rad as a unit is for convenience and a reminder that the angle is not measured in degrees. Radians are not cancelled as meters or seconds or other conventional units. Radian or "rad" is a phantom unit: sometimes it is used and sometimes it is not used. A common expression such as 10 revolutions per minute (rpm) in radians would be $10(2\pi$ rad$)$ per minute.

Example 8-6 Convert 76° to radians and 1 radian to degrees.

Solution: Use unit multiplication here. Watch the units and keep the ratios correct and everything will work out fine.

$$76° \frac{2\pi \text{ rad}}{360°} = 1.33 \text{ rad} \qquad 1 \text{ rad} \frac{360°}{2\pi \text{ rad}} = 57.3°$$

You need to remember that 2π rad = 360°. The other number, 57.3°/radian is not so important and can be worked out with the 2π = 360° definition.

Danger

Go through the following exercise so you are absolutely sure you know how to go back and forth between radians and degrees. This is another mistake that is embarrassing and costly on exams.

Place your calculator in degree mode and take sin 57.3°. You should see 0.84 displayed. Now place your calculator in rad mode. There is usually a DRG (degree, rad, grad) key that cycles through the various modes. There also should be some indicator on the face of the calculator indicating the mode, usually a D or R or G. In rad mode take sin 1. (One radian is approximately 57.3°) You should see the same number, 0.84, displayed. Try a few other angles in radians and degrees to insure that you know how to find the trigonometric function of any angle, whether in degrees or radians.

Refer to Fig. 8-10, which shows a triangle with a very small angle inscribed in a small part of a circle. The angle measured in radians, and the sine and tangent of the angle, are defined in Fig. 8-10.

$$\theta = \frac{s}{c} \qquad \sin\theta = \frac{b}{c} \qquad \tan\theta = \frac{b}{a}$$

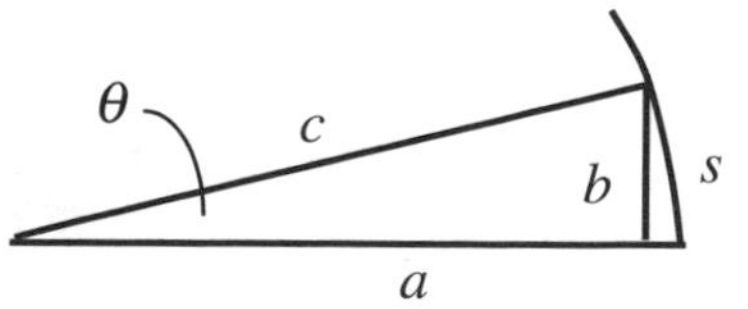

Fig. 8-10

For small enough angles, s is approximately the same as b, and a is approximately the same as c. Therefore, for small angles with the angle measured in radians, the angle, the sine of that angle, and the tangent of that angle are all nearly equal. The next problem illustrates the error in making the approximation that the sine is the angle for some small angles.

Example 8-7   What is the difference (error) between the angle in radians, the sine, and the tangent for an angle of 0.1 radians?

Solution: As a warm up to this problem take the sine of 5.7° and the tangent of 5.7°.

$$\sin 5.7° = 0.0993 \qquad \tan 5.7° = 0.0998$$

The difference between these two is approximately 5 parts in 1000 or 0.5% error. Now

$$\theta \,(\text{in radians}) = 0.1000$$

$$\sin(0.1) = 0.0998$$

$$\tan(0.1) = 0.1003$$

The difference between the sine and the angle at 0.1 radians is 2 parts in 1000.

The difference between the tangent and the angle at 0.1 radians is 3 parts in 1000.

The difference between the sine and the tangent at 0.1 radians is 5 parts in 1000.

Example 8-8   Redo Example 8-7 but at 0.5 radians, approximately 30°.

Solution: $\theta = 0.5$, $\sin\theta = 0.48$, $\tan\theta = 0.55$

The difference between θ and $\sin\theta$ at 0.5 radians is about 4%, and the difference between θ and $\tan\theta$ is about 10%.

Use the approximation $\theta = \sin\theta = \tan\theta$ for angles up to 10° and possibly 20°, but certainly not much bigger. If you find an old slide rule, you will see a scale for sine and tangent that goes to 5.7°

8-4   Non-Right Angle Trigonometry

You may encounter some situations requiring the side or angle in a non-right triangle. The laws relating the sides and angles in non-right triangles are not surprisingly called the *law of sines* and the *law of cosines*. These laws will not be derived; rather they will be stated and illustrated with problems. Actually the law of sines and the law of cosines are applicable to any triangle. Their greatest utility, however, is in non-right triangles.

Law of Cosines

Referring to Fig. 8-11, the law of cosines is written symbolically as $b^2 + c^2 + 2bc(\cos A) = a^2$ Small letters refer to the sides and capital letters to the angles.

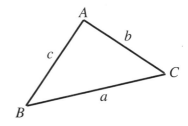

Fig. 8-11

In words, the law of cosines is "one side squared plus an adjacent side squared minus twice the product of the two sides and the cosine of the included angle equals the side opposite the angle squared." This statement is a clearer explanation of the law of cosines. Any side and adjacent side and included angle follow the law of cosines.

Follow the statement, refer to Fig. 8-11, and write the following:

$$b^2 + a^2 - 2ab(\cos C) = c^2 \quad \text{or} \quad c^2 + a^2 - 2ac(\cos B) = b^2$$

Example 8-9 Find the base and the two equal angles of an isosceles triangle with equal sides 4 and included angle, 40°.

Solution: An isosceles triangle has two sides and the opposite two angles equal. The word "base" implies that the unequal side is horizontal. An isosceles triangle as described in the problem statement is shown in Fig. 8-12.

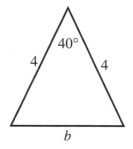

Fig. 8-12

Following the written statement of the law of cosines, "One side squared plus an adjacent side squared minus twice the product of the two sides and the cosine of the included angle equals the side opposite the angle squared," the equation can be written as

$$4^2 + 4^2 - 2 \cdot 4 \cdot 4(\cos 40°) = b^2$$

$$32 - 32(\cos 40°) = b^2$$

$$b^2 = 7.5 \quad \text{or} \quad b = 2.7$$

The base is 2.7, and to complete the picture the equal angles are 70° each
(70° + 70° + 40° = 180°).

Example 8-10 Find the distance and angle to the final position for a person
who travels 6 m at 20° north of east and then 8 m at 50° north of east.

Solution: Instead of an *x-y* coordinate system use the N-S-E-W system representing the compass directions and place arrows representing the 6- and 8-m
distances as shown in Fig. 8-13.

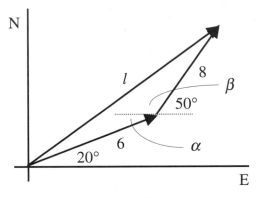

Fig. 8-13

Perhaps the hardest part of this problem is finding the large angle $(\alpha + \beta)$
between the 6- and 8-m lengths. The dashed line at the tip of the line representing
the 6 m is parallel to the E-axis, so the little angle labeled α is 20° (alternate
interior angles of a bisector of parallel lines). The angle labeled β is 130°
(180° − 50°). Therefore, the large angle between the 6- and 8-m lines and opposite the line from the start to the finish is the sum of these two angles
130° + 20° = 150°.

Now write the law of cosines for the length *l*.

$$6^2 + 8^2 - 2 \cdot 6 \cdot 8(\cos 150°) = l^2$$

Before going any farther on this problem, stop and look at the cos 150°. Take
cos 150° on your hand calculator and you will see −0.87 displayed. This is most
reasonable. The length, *l*, is greater than 6 or 8, and just from looking at the
sketch *l* is close to 14, the sum of 6 and 8. With a negative number for the cos 150°,

the law of cosines looks as though it is going to produce a reasonable number. Now proceed with the calculation.

$$36 + 64 - (2)(6)(8)(-0.87) = l^2$$

$$l^2 = 183 \quad \text{or} \quad l = 13.5$$

The important point to notice in this problem, and that is why it was included, is that the law of cosines works for angles greater than 90°.

Law of Sines

Referring to Fig. 8-14, the law of sines is written symbolically as

$$\frac{\sin A}{a} = \frac{\sin B}{b} = \frac{\sin C}{c}$$

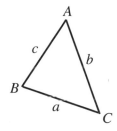

Fig. 8-14

The law of sines works in some instances when the law of cosines does not. The following problem is an example where the law of sines works and the law of cosines does not.

Example 8-11 For a non-right triangle with angles $A = 30°$ and $B = 40°$ and one opposite side $b = 11$, find all the sides and angles.

Solution: Sketch the triangle as in Fig. 8-15.

Notice that the law of cosines will not work in this problem. There are not two sides given.

Using the law of sines $\dfrac{\sin 40°}{11} = \dfrac{\sin 30°}{a}$ and solv-

ing for a we calculate $a = 11\dfrac{\sin 30°}{\sin 40°}$

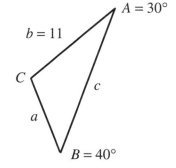

Fig. 8-15

Do not at this point divide 30 by 40; take the sine and multiply by 11!

Carefully find $\sin 30°$, then divide by $\sin 40°$, and finally multiply by 11.

$$a = 11\frac{\sin 30°}{\sin 40°} = 11\frac{0.50}{0.64} = 8.6$$

The angle C is 110° (180° − 30° − 40° = 110°).

And finally use the law of sines to find side c.

$$\frac{\sin 110°}{c} = \frac{\sin 40°}{11} \quad \text{or} \quad c = 11\frac{\sin 110°}{\sin 40°} = 11\frac{0.94}{0.64} = 16.2$$

Example 8-12   You need to determine the distances from two points, A and B, on one side of a canyon to a single point, C, on the opposite side of the canyon. At point A you measure an angle of 65° between points B and C. At point B, 300 m from point A, you measure an angle of 40° between points A and C. What is the distance from A to C and B to C?

Solution: Draw a diagram and use the law of sines. The angle at C is from 180° − 65° − 40° = 75°. Set up the law of sines and note that this law can be written upside down from the normal way (see Fig. 8-16).

$$\frac{300}{\sin 75°} = \frac{a}{\sin 65°} = \frac{b}{\sin 40°}$$

$$a = \frac{\sin 65°}{\sin 75°}\,300 = 281$$

$$b = \frac{\sin 40°}{\sin 75°}\,300 = 200$$

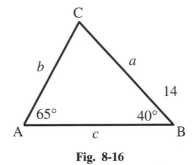

Fig. 8-16

8-5 Trigonometric Functions

The definition of the trigonometric functions starts with the right triangle inscribed in the unit circle first shown in Fig. 8-8.

Place a unit circle on a right angle coordinate system as shown in Fig. 8-17. The sides of the inscribed triangle are the sine and cosine of the angle, θ. These sides of the inscribed triangle are also the projections of the point on the circle that defines the angle on the axes. The projection on the x-axis is the cosine of θ, and the projection on the y-axis is the sine of θ.

As the point defining the angle moves around the circle in a counterclockwise manner, the projection on the x-axis traces out the cosine function and the

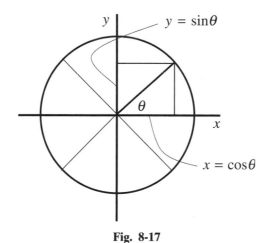

Fig. 8-17

projection on the y-axis traces out the sine function. We will eventually graph the sine function vs. angle, but right now work with the unit circle a little longer.

Follow the sine function and confirm the values in Table 8-1 as the angle is increased. Remember that the angle increases counterclockwise from what would be the +x-axis.

Sine

At $\theta = 0$, the projection on the y-axis that is the value of the sine function is 0.

At $\theta = \pi/4$ or 45°, the projections on the x- and y-axes are the same. Applying the Pythagorean theorem, two equal lengths, l, squared equal the radius (of 1) squared.

$$l^2 + l^2 = 1^2 \quad \text{or} \quad 2l^2 = 1 \quad \text{or} \quad l^2 = 1/2 \quad \text{or} \quad l = 1/\sqrt{2} = 0.71$$

With your hand calculator, confirm that $\sin 45° = \cos(\pi/4) = 0.71$. Confirm this number in Table 8-1.

TABLE 8-1

	0	$\pi/4$	$\pi/2$	$3\pi/4$	π	$5\pi/4$	$3\pi/2$	$7\pi/4$	2π
sin	0	0.71	1	0.71	0	−0.71	−1	−071	0
cos	1	0.71	0	−0.71	−1	−0.71	0	0.71	1
tan	0	1	± ∞	−1	0	1	± ∞	−1	0

At $\theta = \pi/2$, the projection on the y-axis is 1.

At $\theta = 3\pi/4$, the projection on the y-axis is positive and equal to the value at $\pi/4$.

At $\theta = \pi$, the sine function goes to zero.

At $\theta = 5\pi/4$, the projection on the y-axis is negative, but numerically equal to the value at $\pi/4$.

At $\theta = 3\pi/2$, the sine function has value -1, and at $\theta = 7\pi/4$, -0.71, and finally back to zero at 2π.

Cosine

At $\theta = 0$, the cosine function is the projection on the x-axis, or 1. As θ goes from 0 to 2π, the shape of the cosine curve is the same as the shape of the sine curve. They just start at different places; the sine curve starts at zero and the cosine curve starts at 1.

Tangent

The tangent function can be thought of as either y-projection over x-projection or sine function over cosine function. Use whichever is more convenient.

At $\theta = 0$, $\sin\theta = 0$, and $\cos\theta = 1$ so $(\sin\theta/\cos\theta) = 0$.

At $\theta = \pi/4$, the projections are the same so $\tan(\pi/4) = 1$.

At $\theta = \pi/2$, the $\sin(\pi/2)$ over $\cos(\pi/2)$ is 1/0. There is no point at 1/0 so look to a limit view of how the tangent curve behaves in the vicinity of $\pi/2$. As θ approaches $\pi/2$, with values less than $\pi/2$, the $\cos\theta$ becomes small making the tangent of θ become a very large positive number. When θ goes just beyond $\pi/2$, $\cos\theta$ is a small negative number making $\tan\theta$ a very large negative number.

On one side of $\pi/2$ the tangent function goes to plus infinity and on the other side it goes to minus infinity. The best way to depict this on the chart is with $\pm\infty$. In limit language, the tangent function has a vertical asymptote at $\pi/2, 3\pi/2$, and every π interval in both directions. The tangent function is usually graphed between $-\pi/2$ and $\pi/2$ so a complete curve from minus infinity to plus infinity is shown.

The sine, cosine, and tangent functions are graphed in Figs. 8-18 and 8-19.

8-6   Identities

It is not our intention to work out all of the many trigonometric identities. What we will do is show you how broad categories of identities are developed, working

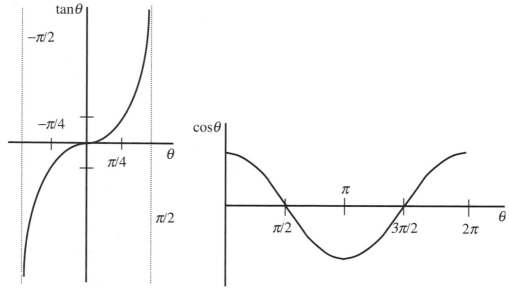

Fig. 8-18

out a few examples along the way. Our purpose is to give you a flavor for trigonometric identities, not make you an expert at them.

The simplest of the identities are the *reciprocals of the sine, cosine,* and *tangent functions.* These are called the cosecant (csc), secant (sec), and cotangent (cot).

$$\csc \theta = \frac{1}{\sin \theta} \quad \sec \theta = \frac{1}{\cos \theta} \quad \cot \theta = \frac{1}{\tan \theta}$$

Look back to Fig. 8-8 and Fig. 8-17 and notice that the inscribed right triangle has radius 1 and write the Pythagorean theorem statement for these inscribed triangles. In terms of the *x* and *y* components, the statement would be

$$x^2 + y^2 = 1$$

In terms of the trigonometric functions the statement would be

$$\sin^2\theta + \cos^2\theta = 1$$

This last statement is often called trigonometric identity number one. Divide this statement by $\cos^2\theta$ to obtain

$$\tan^2\theta + 1 = \frac{1}{\cos^2\theta} \quad \text{or} \quad 1 + \tan^2\theta = \sec^2\theta$$

A variety of similar identities based on $\sin^2\theta + \cos^2\theta = 1$ can be created and are tabulated in the Mathematical Tables.

Another category of identity concerns the *sum or difference of two angles and angles plus or minus* 180°. These can be worked out with the unit circle but they are easier to see from the function graphs. Look just at the sine function graph in Fig. 8-19 and follow the argument presented below.

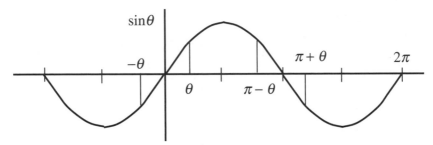

Fig. 8-19

On the sine function graph a vertical line is drawn indicating the position and value of θ. The point $-\theta$ has the same numeric value for the sine function as θ (it is just negative) so identitywise $\sin\theta = -\sin(-\theta)$. Similarly $\sin\theta$ and $\sin(\pi - \theta)$ have the same value, so $\sin\theta = \sin(\pi - \theta)$. The $\sin(\pi + \theta)$ has the same numeric value as $\sin\theta$, one is just the negative of the other, so $\sin\theta = -\sin(\pi + \theta)$. The relations between the sine and cosine are a little more complicated, but not much. Many of the popular trigonometric identities dealing with different angles being equal to or the negative of one another are listed in the Mathematical Tables.

Similar to these relationships are the 90° *difference identities*. These can be worked out geometrically, but are most easily seen in Fig. 8-18. The sine and cosine functions are displaced by 90°. Look at the two graphs and write $\sin\theta = \cos(\theta - 90°)$ and $\cos\theta = \sin(90° + \theta)$. There are more of these in Mathematical Tables.

Another category of identities is the *sum and difference formulas* and the half and double angle formulas. Many of these identities come about from a derivation similar to the one given below for the cosine of the difference of two angles. In addition, this exercise is a good review of basics.

Figure 8-17 shows that any point on the unit circle can be given by the coordinates x and y or the coordinates $\cos\theta$ and $\sin\theta$. On a unit circle the coordinates of the angle θ are $(\cos\theta, \sin\theta)$. The coordinates of another angle, ϕ, are $(\cos\phi, \sin\phi)$ (see Fig. 8-20).

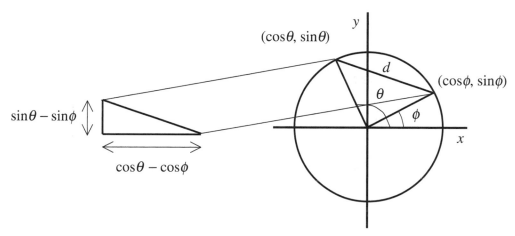

Fig. 8-20

The distance between these two points, d, in terms of the Pythagorean theorem, is

$$d^2 = (\sin\theta - \sin\phi)^2 + (\cos\theta - \cos\phi)^2$$

In words, this is "the hypotenuse d squared equals the difference in x-coordinate squared plus the difference in the y-coordinate squared."

$$d^2 = [\sin^2\theta - 2\sin\theta\sin\phi + \sin^2\phi] + [\cos^2\theta - 2\cos\theta\cos\phi + \cos^2\phi]$$

$\sin^2\theta + \cos^2\theta = 1$ and $\sin^2\phi + \cos^2\phi = 1$ so this statement reduces to $d^2 = 2 - 2\sin\theta\sin\phi - 2\cos\theta\cos\phi$

The distance d^2 can also be written in terms of the law of cosines. The sides are 1, and the included angle is $(\theta - \phi)$, which makes the statement easier.

$$d^2 = 1^2 + 1^2 - 2(1)(1)\cos(\theta - \phi)$$

Set these two statements equal to produce an identity.

$$2 - 2\cos(\theta - \phi) = 2 - 2(\sin\theta\sin\phi + \cos\theta\sin\phi)$$
$$\cos(\theta - \phi) = \sin\theta\sin\phi + \cos\theta\cos\phi$$

This identity gives the cosine of the difference of two angles in terms of the sines and cosines of the individual angles. A myriad of sum and difference of two angle formulas as well as double and half angle formulas come from exercises similar to this one. Fortunately they are all tabulated in many places, most notable in Mathematical Tables in the back of this book. These tables given here are not complete, just sufficient for most of the problems you will encounter.

8-7 Differentiating Trigonometric Functions

A somewhat intuitive justification for the derivative of the sine function was given in Chapter 3, Derivatives. Now it is time to look a little more closely at derivatives of trigonometric functions and apply those derivatives to some problems.

The derivative of the sine function is the cosine function and the derivative of the cosine function is the negative of the sine function. The justification for the derivative of the sine function (Chapter 3) is enough to give you a feel for how the derivatives of trigonometric functions come about. The more popular derivatives are listed below, with a larger list presented in the Mathematical Tables.

$$d(\sin \theta) = \cos \theta d\theta, d(\cos \theta) = -\sin \theta d\theta, d(\tan \theta) = \sec^2 \theta d\theta$$

One thing that occurs fairly often in trigonometric functions is that the variable, the θ, the x, or whatever is not simply θ or x but something more complicated, like $2x$, for example.

Example 8-13   Find the derivative of $\cos 2x$.

Solution: This kind of problem is best done in an implicit style. Write $d(\cos 2x)$ and then differentiate according to the formula for the differential of the cosine.

$$d(\cos 2x) = -(\sin 2x)d(2x) = -2(\sin 2x)dx$$

In words, "the differential of $\cos 2x$ is equal to minus $\sin 2x$ (times the) differential of $2x$." And the differential of $2x$ is $2dx$.

With this experience $\dfrac{d}{dx}(\cos 2x) = -2\sin 2x$.

The implicit derivative approach with the equation written in one line is the easier method.

Example 8-14 Find $\dfrac{dy}{dx}$ for $y = \sin^3 x$.

Solution: Before differentiating the $\sin x$ we first have to deal with the cube of $\sin x$. If the problem were $y = u^3$, the implicit style derivative would read $dy = 3u^2 du$.

For the function $y = \sin^3 x$, an approach parallel to the one above would produce $dy = 3(\sin^2 x)d(\sin x)$.

The $d(\sin x)$ is $(\cos x)dx$, so $\dfrac{dy}{dx} = 3(\sin^2 x)(\cos x)$.

Example 8-15 Find $\dfrac{dy}{dx}$ of $y = \tan^2 2\theta$.

Solution: Go slowly and don't get confused. Do the tangent squared part of the derivative and then take care of the 2θ part.

$$dy = 2(\tan 2\theta)d(\tan 2\theta)$$

$$dy = 2(\tan 2\theta)(\sec^2 2\theta)d(2\theta)$$

$$d = 4(\tan 2\theta)(\sec^2 2\theta)d\theta \quad \text{or} \quad \frac{dy}{d\theta} = 4(\tan 2\theta)(\sec^2 2\theta)$$

The problem can be rewritten using the identity $\sec^2 2\theta = 1 + \tan^2 2\theta$. Rewrite the problem as

$$y = \sec^2 2\theta - 1$$

Take the differential in implicit form

$$dy = 2\sec 2\theta\, d\sec 2\theta = 2\sec 2\theta(\tan 2\theta \sec 2\theta d2\theta) = 4\tan 2\theta \sec^2 2\theta d\theta$$

so

$$\frac{dy}{d\theta} = 4\tan 2\theta \sec^2 2\theta$$

Example 8-16 What is the angle between $y = \cos 3x$ and the x-axis when the curve crosses the axis?

Solution: This is one of those innocent looking problems that look easy but perhaps is not so easy. Oh, but it is not difficult for Captain Calculus, because "The Captain" always "thinks calculus."

The phrase "the angle" should trigger a connection between geometry and calculus. To know the slope is to know the angle so if we know the slope when the curve crosses the axis then we can easily find the angle. The slope, in general, is the derivative.

$$dy = -\sin 3x\, d3x = -3\sin 3x\, dx$$

or $\quad \dfrac{dy}{dx} = -3\sin 3x$

What we need is the specific slope when the function $y = \cos 3x$ crosses the x-axis. We have the general expression for the slope so all we need is the x-value when the function crosses the x-axis.

The function $y = \cos 3x$ crosses the x-axis when $3x = \pi/2$ or $x = \pi/6$. Substituting $\pi/6$ into the general expression for the slope, we calculate

$$\left.\frac{dy}{dx}\right|_{x=\pi/6} = -3\sin(3\pi/6) = -3\sin(\pi/2) = -3$$

Figure 8-21 shows the first quarter cycle of the curve $y = \cos 3x$. The slope is clearly negative at this point. The angle between the axis and the curve as shown in Fig. 8-21 has tangent of 3. Solve $\tan^{-1}\theta = 3$ for $\theta = 72°$.

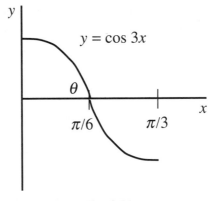

Fig. 8-21

Example 8-17 The rise and fall of ocean tides follows $y = (3\,\text{ft})\sin\!\left(\dfrac{2\pi}{11}t\right)$, where y is the relative height of the ocean, taking $y = 0$ as the midpoint between high and low tide, and t is the time in hours from the midpoint in height. When, in the cycle, is the tide rising at its greatest rate, what is that rate?

Solution: The sine function describes the up and down motion of the tide. The 3 ft is the height or depth of the ocean from the midpoint between high and low tide.

The $2\pi/11$ is determined by the frequency of the tide and the nature of the sine function. The time for one tide cycle is approximately 11hr. When t has gone from 0 to 11, the argument of the sine function, the $(2\pi/11)t$, has gone from 0 to 2π, or through one complete cycle. Read this paragraph again and perhaps again, until you understand how to write descriptions of processes that vary in a sinusoidal manner.

The function is graphed in Fig. 8-22. The vertical scale shows the 3 ft up and down of the tide and the horizontal scale shows one complete cycle after 11 hr.

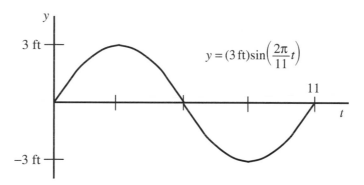

$$y = (3\,\text{ft})\sin\!\left(\frac{2\pi}{11}t\right)$$

Fig. 8-22

The rate at which the tide is rising is the time derivative of y.

$$\frac{dy}{dt} = (3\text{ft})\left(\frac{2\pi}{11\ \text{hr}}\right)\cos\left(\frac{2\pi}{11}t\right) = \left(\frac{6\pi\ \text{ft}}{11\ \text{hr}}\right)\cos\left(\frac{2\pi}{11}t\right)$$

The $2\pi/11$ has the units 1/hr. If t is measured in hours then $2\pi/11$ must have the units of reciprocal hours so that the sine is of a pure number. It is impossible to take the sine of 3 ft or 6 hr or \$1.25. The only choices in taking a sine are a pure number (radians) or degrees. Use this pure number requirement to keep units straight.

The rate at which the tide is rising, dy/dt, is a maximum when $\cos(2\pi/11)t$ is maximum. The cosine function is a maximum at 0 or in this case at $t = 0$. This point, $t = 0$, corresponds to the midpoint between high and low tide (see Fig. 8-21).

The tide is rising at its fastest rate midway between high and low tide.

The maximum rate is $\dfrac{2\pi}{11}\dfrac{\text{ft}}{\text{hr}} = 0.57\dfrac{\text{ft}}{\text{hr}}$.

Second Solution: Captain Calculus, who always "thinks calculus," would not need to take a derivative to know when the tide was rising at maximum rate. The Captain would look at the sine curve (Fig. 8-22) describing how the ocean level was going up and down with the tide and ask where the slope had the greatest positive value. Just by looking at the curve, the maximum positive slope and the greatest rate of rise of the tide are at the midpoint between high and low tide.

8-8   Integrating Trigonometric Functions

Three basic integral formulas can be obtained by taking the antiderivative of the differential formulas at the beginning of the previous section.

$$\int d(\sin\theta) = \int \cos\theta d\theta, \ \int d(\cos\theta) = -\int \sin\theta d\theta, \ \int d(\tan\theta) = \int \sec^2\theta d\theta$$

$$\int \cos\theta d\theta = \sin\theta, \ \int \sin\theta d\theta = -\cos\theta, \ \int \sec^2\theta d\theta = \tan\theta$$

Two other popular integrals of trigonometric functions are

$$\int \tan\theta d\theta = -\ln(\cos\theta), \ \int \cot\theta d\theta = \ln(\sin\theta)$$

These and a few other trigonometric integrals are listed at the back of the book in the Mathematical Tables.

Handling integrals other than the standard integral, $\int \sin$, $\cos$, or $\tan(x)dx$, is a little harder with integrals than it was with differentials. The following problem illustrates the procedure.

Example 8-18 Find the integral of $y = \int \sin 2\theta d\theta$.

Solution: Here again is a case where $\int \sin u du$ is known but this is not the exact problem. The first step in this problem is to make the $\int \sin 2\theta d\theta$ look like $\int \sin u du$.

Constants can be placed inside or outside the integral sign; it makes no difference. To make this problem read $\int \sin u du$, multiply by 2/2 and take the 2 in the numerator inside the integral and associate it with the $d\theta$, and leave the remaining 1/2 outside the integral.

$$y = \frac{1}{2} \int \sin 2\theta d2\theta$$

Now the integral is in the form $\int \sin u du$ and has the following solution:

$$y = \frac{1}{2}(-\cos 2\theta) = -\frac{1}{2}\cos 2\theta$$

Example 8-19 Find $y = \int \sin^3 \theta d\theta$.

Solution: Don't be fooled by this integral. It is not a power law problem. It is not an easy integral and do not get involved in trying to work it out. Go to the Mathematical Tables or some table of integrals and copy the answer.

$$y = \int \sin^3 \theta d\theta = -\frac{1}{3}\cos\theta(\sin^2\theta + 2)$$

Example 9-20 Find the average value of the sine function from 0 to π.

Solution: The average value of the function uses the definition of the average value of a function over a range (see Chapter 7, Integration). The integration is over the first half-cycle of the sine function as shown in Fig. 8-22. The average value of the function $x = \sin\theta$ from 0 to π is

$$x|_{avg} = \frac{1}{\pi - 0}\int_0^\pi \sin\theta d\theta$$

Carry out the integration to find the average value of the sine function over one half-cycle.

$$x|_{avg} = \frac{1}{\pi}\left[-\cos\theta\right]_0^\pi = \frac{1}{\pi}[-\cos\pi - (-\cos 0)]$$

$$= \frac{1}{\pi}[-(-1) - (-1)] = \frac{2}{\pi} = 0.64$$

Figure 8-23 shows the rectangle with height 0.64 and base π with area equal to the area under the first half-cycle of $\sin\theta$.

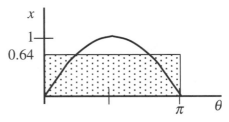

Fig. 8-23

Example 8-21   The power delivered by a loudspeaker is $P = P_o\sin^2\omega t$ where P_o is the peak power and ω is a constant with the units of reciprocal time. What is the average power in terms of the peak power?

Solution: Start by graphing $\sin\omega t$ (Fig. 8-24). The ωt is not important to the graph. When ωt has gone from 0 to 2π, the sine function has gone through one cycle.

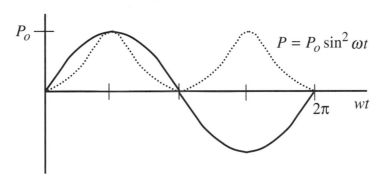

Fig. 8-24

Now graph the $\sin^2$ curve. Look first at the range from 0 to π. The $\sin^2$ curve starts at zero when $\omega t = 0$, and goes to 1 when $\omega t = \pi/2$, and then back to zero again when $\omega t = \pi$.

The $\sin^2$ curve, however, has a different shape from the sin curve. The unique shape of the $\sin^2$ curve is due to the fact that when a number less than 1 is squared, the result is smaller [$0.5^2 = 0.25$]. The smaller the number, the smaller the result on squaring [$0.9^2 = 0.81$ but $0.3^2 = 0.09$].

When the sin curve is negative, the $\sin^2$ curve is positive (see Fig. 8-24). The $\sin^2$ curve is periodic in π so the average value of the $\sin^2$ curve is the average value between 0 and π.

The average value of this $\sin^2$-type function follows the definition of the average value of the function.

$$P|_{avg} = \frac{1}{\pi - 0} P_o \int_0^\pi \sin^2 \omega t \, d(\omega t)$$

The $\int \sin^2 \theta \, d\theta$ integral is (from the Mathematical Tables) $\frac{1}{2}\theta - \frac{1}{4}\sin 2\theta$ so

$$P_{avg} = \frac{P_o}{\pi}\left[\frac{1}{2}\omega t - \frac{1}{4}\sin 2\omega t\right] = \frac{P_o}{\pi}\left\{\left[\frac{\pi}{2} - \frac{1}{4}\sin 2\pi\right] - \left[0 - \frac{1}{4}\sin 0\right]\right\} = \frac{P_o}{2}$$

The average power for the loudspeaker is one-half the peak power.

Example 8-22 What is the area bounded by $y_1 = \cos x$, $y_2 = \sin x$, and $x = 0$?

Solution: The sine and cosine functions are shown in Fig. 8-25. The integral of the area between the curves is in the x-direction and has form $\int(\cos x - \sin x)dx$. The integral is from $x = 0$ to the intersection point of the two curves. At this point

$\sin x = \cos x$ or $\frac{\sin x}{\cos x} = 1$ or $\tan x = 1$

From the graph of $\tan x$, $\tan x = 1$ when $x = \pi/4$. Check the number in your hand calculator. Take the inverse tangent of 1. Table 8-1 also shows $\sin x$ equal to $\cos x$ at $x = \pi/4$ so the complete integral, complete with limits, is

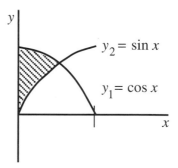

Fig. 8-25

$$A = \int_0^{\pi/4}(\cos x - \sin x)dx$$

$$A = [\sin x + \cos x]_0^{\pi/4} = \left[\sin\frac{\pi}{4} + \cos\frac{\pi}{4}\right] - [\sin 0 + \cos 0]$$

$$A = 0.71 + 0.71 - 1 = 1.41 - 1 = 0.41$$

Example 8-23 A certain company is selling golf shirts with the college logo through college bookstores. Because of an expanding market overall sales are growing at $2t$ thousands of dollars per month, t. Superimposed over this linear growth is a sinusoidal variation following the academic year and described by $(0.1)\cos 2\pi\frac{t}{12}$ in thousands of dollars and with $t = 0$ corresponding to January. Determine the rate of sales at the end of March ($t = 3$).

Solution: The equation describing sales is

$$S = 2t + 0.1\cos 2\pi\frac{t}{12}$$

The general expression for the change in sales is

$$dS = 2dt - (0.1)\sin\frac{2\pi t}{12}d\left(\frac{2\pi t}{12}\right) \quad \text{or} \quad \frac{dS}{dt} = 2 - \frac{(0.1)\pi}{6}\sin\frac{2\pi t}{12}$$

The rate of change of sales at the end of March ($t = 3$) is obtained by evaluating the $\frac{dS}{dt}$ at $t = 3$.

$$\left.\frac{dS}{dt}\right|_{t=3} = 2 - \frac{(0.1)\pi}{6}\sin\frac{\pi}{2} = 2$$

At the end of March sales are at the linear rate of 2 thousand dollars per month.

Example 8-24 For the previous problem find the total sales for the 3 months from January through March.

Solution: Sales follow $S = 2t + 0.1\cos 2\pi\frac{t}{12}$. The total sales over any period is the integral of S, with appropriate limits. The limits for this January to March period are $t = 0$ and $t = 3$.

$$S_{0\to3} = \int_0^3 [2t + (0.1)\cos \pi t/6]dt$$

$$S_{0\to3} = \int_0^3 2t\,dt + \int_0^3 0.1\frac{6}{\pi}\cos\frac{\pi t}{6}\,d\frac{\pi t}{6}$$

$$S_{0\to3} = t^2\Big|_0^3 + \frac{0.6}{\pi}\sin\frac{\pi t}{6}\Big|_0^3$$

$$S_{0\to3} = 9 + \frac{0.6}{\pi}\left[\sin\frac{\pi}{2} - \sin 0\right]$$

$$S_{0\to3} = 9 + \frac{0.6}{\pi} = 9.2$$

Total sales for the January to March period were 9.2 thousands of dollars.

It's a Wrap

✔ Solve right and non-right triangles

✔ Know how to derive simple identities

✔ Differentiate trigonometric functions

✔ Integrate trigonometric functions

✔ Apply differentiation and integration to practical problems

✔ Use tables for differentiation of trigonometric functions

✔ Use tables for integration of trigonometric functions

Test Yourself

PROBLEMS

1. A cross country skier skis 1.2 km north and then 2.0 km east. How far and in what direction is she from her starting point?
2. Two sides of a non-right triangle are 4.0 and 5.0 and the included angle is 38°. What are the other angles and side?
3. Convert: 2 rad to degrees, $\pi/3$ to degrees and 85° to radians.
4. Find the derivative of $f(s) = \sin(3s + 1)$.
5. Find the derivative of $f(t) = \cos^2 t$.
6. $y(x) = \tan(x^2) + \sec(2 - 3x)$.
7. Evaluate $\int \sec x(\sec x - \tan x)dx$.
8. Evaluate $\int x \sin x^2 dx$.

9. Sales of a certain product follow a cyclical sales pattern according to $S(t) = 30 + 12\cos\dfrac{\pi t}{8}$, where t is the time in months measured from January 1 and S is the number of sales in thousands. What is the average sales for the first quarter of the year?

10. What is the area under the curve of $x \sin x$ from 0 to 2π.

ANSWERS

You may need to consult a table of differentials and integrals to solve these problems.

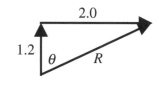

1. Draw a vector diagram representing her trip. Her <u>distance from</u> the starting point is $R = \sqrt{1.2^2 + 2.0^2} = 2.3$ km

   The angle is $\theta = \tan^{-1}\dfrac{2}{1.2} = 59°$

2. Use the law of cosines to find side b.
   $b^2 = 4^2 + 5^2 - 2(4)(5)\cos 38°$
   so $b = 3.08$
   Use the law of sines to find A and C.

   $\dfrac{\sin A}{5.0} = \dfrac{\sin 38°}{3.08} = \dfrac{\sin C}{4.0}$

   $\sin A = \dfrac{5}{3.08}\sin 38° \quad A = 88°$

   $\sin C = \dfrac{4}{3.08}\sin 38° \quad C = 53°$

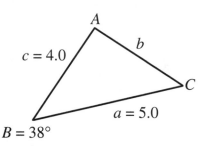

3. $2\,\text{rad}\,\dfrac{360°}{2\pi\,\text{rad}} = 115° \quad \dfrac{\pi}{3}\dfrac{360°}{2\pi} = 60° \quad 85°\dfrac{2\pi}{360°} = 1.48$

4. Use the chain rule and think of $(3s + 1)$ as another variable, u.
   $df = \cos(3s + 1)d(3s + 1) = 3\cos(3s + 1)ds \quad \Rightarrow \quad f'(s) = 3\cos(3s + 1)$

5. $f'(t) = -2\sin t \cos t$

6. $dy = \sec^2(x^2)d(x^2) + \sec(2 - 3x)\tan(2 - 3x)d(2 - 3x)$
   $\dfrac{dy}{dx} = 2x\sec^2(x^2) - 3\sec(2 - 3x)\tan(2 - 3x)$

7. $\int \sec^2 x\,dx - \int \sec x \tan x\,dx$
   $\tan x - \sec x + C$

8. Make a change of variable: let $u = x^2$ so $du = 2x\,dx$. If there were no x outside the argument of the sin then this would be a much more difficult integral.
   $\dfrac{1}{2}\int \sin u\,du = -\dfrac{1}{2}\cos u$ so integral is $-\dfrac{1}{2}\cos x^2$

9. This is an average value integral over the first 3 months of the year.

$$Avg\, Sales = \frac{1}{3-0}\int_0^3 \left(30 + 12\cos\frac{\pi t}{8}\right)dt = \frac{1}{3}\left[30\int_0^3 dt + 12\int_0^3 \cos\frac{\pi t}{8}dt\right]$$

$$Avg\, Sales = \frac{1}{3}\left[30\int_0^3 dt + 12\frac{8}{\pi}\int_0^3 \cos\frac{\pi t}{8}\; d\frac{\pi t}{8}\right]$$

$$Avg\, Sales = \frac{1}{3}\left[30t + \frac{96}{\pi}\sin\frac{\pi t}{8}\right]_0^3 = \frac{1}{3}\left[90 + \frac{96}{\pi}\sin\frac{3\pi}{8}\right] = 39$$

The average monthly sales for the first 3 months is 39 thousand.

10. The curve is a somewhat distorted sin function. The area is the definite integral of the function over the interval.

$A = \int_0^\pi x\sin x\, dx$. This integral can be performed by a technique called integration by parts which is taken up in the next chapter or it can be "guessed" with just a few tries. Follow along on the guessing approach. First a cosine is involved because of the integral of the sine, so start with $\cos x$ and differentiate to find the integrand. The differential of $\cos x$ is $-\sin x$ which is a start. Now for a little better guess try $-x\cos x$ which has differential $x\sin x$ which is the integrand but it has the added differential term $-\cos x$. This is taken care of by adding a $\sin x$. The final integrand is from $d(-x\cos x + \sin x)$ $= +x\sin x - \cos x + \cos x = x\sin x$. A little tortured, but not impossible. With a little experience these can go rather quickly. This, however, is probably not high on your list of skills to learn. Buy mathematical tables.

CHAPTER 9

EXPONENTS AND LOGARITHMS

Do I Need to Read This Chapter? You should read this chapter if you need to review or you need to learn about

→ Basics of exponents and logarithms

→ Exponential functions

→ The number e

→ Growth and decay problems

→ Limited growth exponentials

→ The logistic function

The short review of exponents and logarithms in Chapter1, Mathematical Background, presumed a rudimentary knowledge of exponents and logarithms. No such presumption is made in this chapter. Here we start with basic definitions and work up to differentiation and integration of exponential and logarithmic functions.

Exponential functions describe a wide variety of phenomena including radioactive decay, bacteria growth, learning retention, growth of investments, proliferation of disease, certain electrical phenomena, logistic-type growth and on and on, providing many good examples of the application of exponential functions.

When I'm on my birdhouse, I'm as powerful as an exponent!

The statement of some of these phenomena is often quite simple but the specific laws governing them and the predictive ability of these laws require a good understanding of exponents, logarithms, and calculus. This chapter is very applications oriented. No matter what your field of interest, there will be some applications that bear directly on your area of interest.

9-1   Exponent Basics

A number written as 2^3, which is just a short-hand way of writing $2 \cdot 2 \cdot 2$, is a number, 8 in this case, written in exponential form where 2 is called the base and 3 the exponent. Two numbers such as 3^2 and 5^2 cannot be added and the answer written in a meaningful exponential form. The 3^2 is equal to $3 \cdot 3$ and the 5^3 equal to $5 \cdot 5 \cdot 5$. There is no combination of 3s and 2s and 5s that represents the addition of the two numbers. The only way to add the numbers is to write 3^2 as 9 and 5^3 as 125 and add them to obtain 134. Likewise, there is no way to subtract numbers written as exponents. Even 3^2 plus 3^2 cannot be written as an exponent.

Multiplying is much easier: $3^4 \cdot 3^3$ is 3^7, just add the exponents. Visualize 3^4 as four 3s multiplied together and 3^3 as three 3s multiplied together and all of them multiplied together as seven 3s multiplied together.

Dividing is equally easy: $3^5/3^3 = 3^2$. Five 3s divided by three 3s means there are two of them remaining in the numerator.

Raising to a power is a slight variation on multiplying. The form $(2^3)^2$ is viewed as 2^3 times 2^3 or 2^6. These examples illustrate the three basic laws of exponents.

$$a^m \cdot a^n = a^{m+n} \qquad \frac{a^m}{a^n} = a^{m-n} \qquad (a^m)^n = a^{m \cdot n}$$

Negative exponents mean reciprocal or one over: $a^{-m} = \dfrac{1}{a^m}$.

The laws of exponents work equally well for negative and fractional exponents.

Example 9-1 Evaluate: $2^3 \cdot 2^5$; $5^7/5^{11}$; $3^{-7} \cdot 3^4$; $(4^2)^{-6}$; $(7^{3/2})^2$, $3^{0.2}/3^2$.

Solution: For $2^3 \cdot 2^5$ add the exponents to obtain 2^8.

For $5^7/5^{11}$ subtract the exponents keeping the signs correct to obtain $5^{7-11} = 5^{-4}$.

For $3^{-7} \cdot 3^4$ add the exponents keeping the signs correct to obtain $3^{-7+4} = 3^{-3}$.

For $(4^2)^{-6}$ multiply the exponents to obtain 4^{-12}.

For $(7^{3/2})^2$ multiply the exponents keeping the fractions correct to obtain 7^3.

For $3^{0.2}/3^2$ read the problem as $3^{0.2} \cdot 3^{-2}$ and add the exponents to obtain $3^{-2.8}$.

9-2 Exponential Functions

Exponential functions are in the form $y = a^x$. Taking $a = 2$ the function reads $y = 2^x$. This is a rapidly increasing function as tabulated and shown in Fig. 9-1.

The function $y = 2^{-x}$ is also interesting and is tabulated and graphed in Fig. 9-2. Remember that any number raised to the zeroth power is 1.

Example 9-2 Certain cells grow by splitting; one cell begets two and each of these begets two (more) with each cycle taking 3 hr. A simple model for the growth in the number of cells is $N = N_o 2^{t/3}$, where N_o is the number of cells at $t = 0$ and t is the time in hours. If 1000 cells are left to grow over 60 hr, how many cells are there at the end of the 60 hr?

x	0	1	2	3	−1	−2
y	1	2	4	8	1/2	1/4

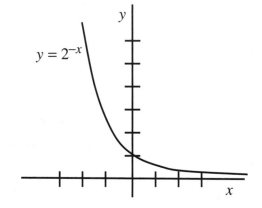

Fig. 9-1

x	0	1	2	−1	−2
y	1	1/2	1/4	2	4

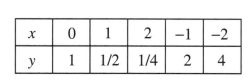

Fig. 9-2

Solution: At time zero there are 1000 cells. At the end of 3 hr there are 2000 cells, and at the end of 6 hr there is another doubling to 4000 cells. The model as described in words and by the equation $N = (1000)2^{t/3}$ is consistent. Make the time go on for 9 hr and the number doubles again to 8000 cells.

Use the formula for this specific situation to find the number at 60 hr.

$$N = N_o 2^{t/3} = (1000)2^{60/3} = (1000)2^{20} = 1.0 \times 10^9 \text{ cells}$$

The power of an exponential function to generate large numbers is tremendous.

Example 9-3 The return on a certain investment follows $R(t) = 1000e^{-0.8t}$, where R is in dollars when t is in months. How fast is the return declining at the third month?

Solution: The question concerns a rate and in calculus talk this means the differential of R with respect to time. So,

$$dR = 1000e^{-0.8t}d(-0.8t) = -800e^{-0.8t}dt \quad \Rightarrow \quad \frac{dR}{dt} = -800e^{-0.8t}$$

The decline in return at 3 months is: $\left.\dfrac{dR}{dt}\right|_{t=3} = -800e^{-0.8(3)} = (-)\$72/\text{month}.$

Example 9-4 The return on \$1000 invested in a mining operation produces the following returns in dollars for the months indicated: $R(1) = 200$, $R(2) = 165$, $R(3) = 145$, $R(4) = 116$, $R(5) = 86$, $R(6) = 72$, $R(7) = 52$, $R(8) = 50$, and $R(9) = 38$. Based on this data, what is the rate the return is declining at the third and sixth months?

Solution: An algebra oriented person would probably take the data for the second and fourth months, find the slope symmetric about the third month and use this as the rate of decline. Captain Calculus would not be content with this approach. The Captain would place the data on a spread sheet, have the spread sheet draw the graph depicting the data and calculate the equation describing the return over time, and then differentiate this equation and evaluate it at month three.

Take me to your spread sheet!

The rate equation (from the spread sheet calculation) is

$$R = 1.8t^2 - 38t + 238$$

and the derivative is

$$\frac{dR}{dt} = 3.6t - 38$$

At month three the rate of decline is

$$\left.\frac{dR}{dt}\right|_{t=3} = 3.6(3) - 38 = -\$27.20/\text{month}$$

This answer is most reasonable. At low t the decline is steep and as time goes on the rate of decline becomes less and less.

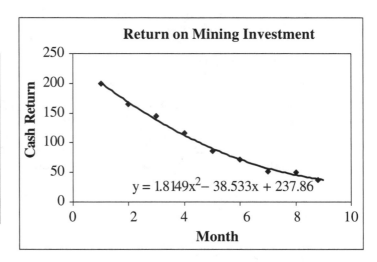

R	t
1	200
2	165
3	145
4	116
5	86
6	72
7	52
8	50
9	38

Example 9-5   The number of hot tubs sold in the United States between 1995 and 2003 followed the model: $n(t) = 0.04t^2 + 0.3t + 15$, where t is the year and $n(t)$ is in thousands of hot tubs. The year 1995 corresponds to $t = 1$ and so on up to $t = 9$ for 2003. What was the rate of hot tub sales in the year 2000?

Solution: The rate of hot tub sales is dn/dt and in the year 2000 the rate was dn/dt evaluated for $t = 6$.

$$\frac{dn}{dt} = 0.08t + 0.3$$

$$\left.\frac{dn}{dt}\right|_{t=6} = 0.08(6) + 0.3 = 0.78$$

According to this model hot tub sales in 2000 were 780 units.

Example 9-6   The return on the purchase of a Wi-Fi tower is predicted to be $R(t) = 2000e^{0.1t}$ over the next 4 years. In this model year 0 is the first year and so on up to the fourth year at $t = 4$. What will be the total return over these 4 years?

Solution: Because of the way the model is written the integral is from 0 to 4. Visualize the area as representing the return for each year summed over the 4 years.

$$\int R(t) = 2000 \int_0^4 e^{0.1t}dt = \frac{2000}{0.1} \int_0^4 e^{0.1t}d(0.1t) = 20,000e^{0.1t}\Big|_0^4$$

$$\int R(t) = 20,000[e^{0.4} - 1] = \$9836$$

The total return for the 4 years when this model is presumed to be valid is $9836.

Example 9-7   The cost in dollars for fuel to operate a truck for t months follows the model: $C(t) = 400(2.20 + 0.1t)$. The 2.20 represents the base cost per gallon of fuel with the $0.1t$ the predicted increase. The 400 is 20 working days per month times 20 gallons per day average. What is the total cost for 1 year (0 to 12 months)?

Solution: Integrate this cost equation over t and from 0 to 12.

$$\int C(t) = 400 \int_0^{12} (2.20 + 0.1t)dt = 400\left[2.20t\Big|_0^{12} + 0.1\frac{t^2}{2}\Big|_0^{12} \right]$$

$$\int C(t) = 400[264 + 7.2] = \$13440/\text{year}$$

The operating cost for the year is $13440.

Example 9-8   A nominal $10,000 invested in a gas well produces the following results for the first 6 months: $R(1) = 260$, $R(2) = 225$, $R(3) = 206$, $R(4) = 195$, $R(5) = 180$, and $R(6) = 178$. Find the equation that fits this data and predict the total return for this investment for the first 12 months.

Place the data in a spread sheet and have the spread sheet program graphically display the data. Then have the program, draw a second degree polynomial approximation and write an equation to fit the data. $R = 3.2t^2 - 38.5t + 293$. Assuming that the model derived from this data will continue for 12 months the predicted value for each month can be determined. The question, however,

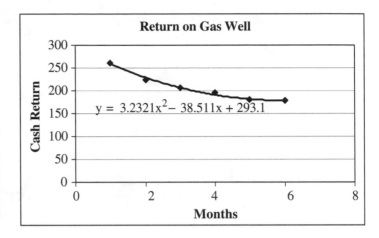

R	t
1	260
2	225
3	206
4	195
5	180
6	178

concerns the total return over the 12 months and this is the integral of this function from 0 to 12.

$$R(12)_{total} = \int_0^{12} (3.2t^2 - 38.5t + 293)dt = \left[\frac{3.2t^3}{3} - \frac{38.5t^2}{2} + 293t\right]_0^{12}$$

$$R(12)_{total} = [1152 - 2772 + 3516] = \$1896$$

With data for 6 months a model was developed using a spread sheet program and this model was used to predict the total return for the first 12 months.

9-3 The Number e

The number e, approximately 2.72, is an irrational number (irrational numbers cannot be written as fractions) that occurs in nature in many different places. Two of the definitions are associated with calculus and are outlined here.

The first definition of e involves a limit. The number e is defined as

$$e = \lim_{x\to\infty} \left(1 + \frac{1}{x}\right)^x$$

That is a rather strange looking definition but one that is easy to check on your hand calculator.

The definition states that as x gets closer and closer to infinity, the operation $\left(1 + \frac{1}{x}\right)^x$ approaches a limit, the number e.

Try a few numbers in your hand calculator. These few short calculations produce a value for *e* good to three significant figures. Your hand calculator probably computes *e* with this algorithm and a high enough value for *x* to reproduce the precision appropriate to your calculator.

x	$1 + \dfrac{1}{x}$	$\left(1 + \dfrac{1}{x}\right)^x$
1	1+1	2
2	1+0.5	2.25
10	1.1	2.59
100	1.01	2.70
1000	1.001	2.72

The second definition of *e* is that $y = e^x$ is the exponential function whose derivative is everywhere equal to the value of the function. In calculus language this means we are looking for a number *a* such that

$$\frac{d}{dx} \text{ of } a^x \text{ is } a^x.$$ Such a number exists and it is the number *e*.

The value of *e* can be determined by taking the function $y = a^x$ and looking for the value of *a* where the slope is equal to the value. Use a simple calculus method for determining the slope as illustrated in Fig. 9-3.

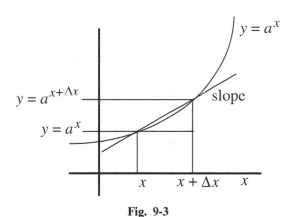

Fig. 9-3

Set up a spread sheet to calculate the slope for various combinations of *a*s. Start with $a = 2$ and $a = 3$. Use the point 2.5 and calculate the value of the function and the slope. Notice that for $a = 2$ the value is higher than the slope and that for $a = 3$ the value is less than the slope. This means that the number whose slope is the same as its value is between 2 and 3.

Number	Exponent	Δx	Value at $x + \Delta x$	Value at x	Slope
a	x	Δx	$a^{x+\Delta x}$	a^x	$(a^{x+\Delta x} - a^x)/\Delta x$
2	2.5	0.1	6.06	5.66	4.06
3	2.5	0.1	17.40	15.59	18.10

Now use a values of 2.6 and 2.8. The slope is lower than the value for $a = 2.6$ and higher for $a = 2.8$. This means that the number is between 2.6 and 2.8.

Number	Exponent	Δx	Value at $x + \Delta x$	Value at x	Slope
a	x	Δx	$a^{x+\Delta x}$	a^x	$(a^{x+\Delta x} - a^x)/\Delta x$
2.6	2.5	0.01	11.005	10.900	10.465
2.8	2.5	0.01	13.255	13.119	13.577

Now use a values of 2.7 and 2.75. This again indicates that the number is between 2.7 and 2.75.

Number	Exponent	Δx	Value at $x + \Delta x$	Value at x	Slope
a	x	Δx	$a^{x+\Delta x}$	a^x	$(a^{x+\Delta x} - a^x)/\Delta x$
2.7	2.5	0.001	11.9906	11.9787	11.9038
2.75	2.5	0.001	12.5537	12.5410	12.6929

Finally take a values of 2.71 and 2.72, which shows that the number e is between 2.71 and 2.72.

Number	Exponent	Δx	Value at $x + \Delta x$	Value at x	Slope
a	x	Δx	$a^{x+\Delta x}$	a^x	$(a^{x+\Delta x} - a^x)/\Delta x$
2.71	2.5	0.0001	12.09112	12.08991	12.05362
2.72	2.5	0.0001	12.20297	12.20175	12.21007

This process can be carried on to the precision of the spread sheet.

The result of this exercise is that the derivative (slope) of the function $y = e^x$ is e^x.

In calculus language this means that $\dfrac{d}{dx}(e^x) = e^x$ or $d(e^x) = e^x dx$.

Likewise the integral of e^u is also e^u. $\displaystyle\int e^u du = e^u$.

These two very calculus oriented questions:

"What number do you get when you take $\lim_{x \to \infty}\left(1 + \frac{1}{x}\right)^x$?" and

"What value of a in the function $y = a^x$ gives a derivative equal to itself?" produce the number e.

Most of the exponential problems in the remainder of this chapter will use the number e. As we get further into the study of logarithms, e will return again as an important number.

Taking this one step further, any function represented by the symbol u is differentiated as

$$d(e^u) = e^u du$$

The integral of e^u is also e^u

$$\int e^u du - e^u$$

Example 9-9 Find $\dfrac{dy}{dt}$ for $y = e^{at}$.

Solution: The safest way to do this problem is in an implicit derivative format.

$$dy = e^{at}d(at) = a(e^{at})dt \text{ and } \frac{dy}{dt} = a(e^{at})$$

Example 9-10 Find $\dfrac{dy}{dt}$ for $y = e^{t^3}$.

Solution: Again, use an implicit derivative format.

$$dy = e^{t^3}d(t^3) = 3t^2(e^{t^3})dt \text{ and } \frac{dy}{dt} = 3t^2(e^{t^3})$$

Example 9-11 Find $\int e^{3x}dx$.

Solution: Change the integral to $\dfrac{1}{3}\int e^{3x}d(3x)$ so it is in standard form, $\int e^u du = e^u$, and write

$$\int e^{3x}dx = \frac{1}{3}\int e^{3x}d(3x) = \frac{1}{3}e^{3x}$$

Example 9-12   What is the accumulated balance on $1000 placed at 6% interest for 5 years if the interest is compounded (a) quarterly or (b) monthly?

Solution: If the interest on a principal amount is compounded once at the end of an interval the amount is $A = P(1 + r)$, where r is the rate of return written as a decimal.

A one time 10% interest payment on $1000 would produce $A = \$1000(1 + 0.10) = \$1100$. If this $1100 remained at the 10% and the interest compounded again at the end of the next interval the amount would be $A = [P(1 + r)](1 + r) = \$1100(1 + 0.10) = \$1210$. The expression in brackets represents the amount after one compounding and the entire expression represents the amount after two compoundings.

Depending on the number of compoundings, in general the amount would be $A = P(1 + r)^n$, where r is the rate for the compounding interval and n is the number of intervals.

Interest is usually stated on a yearly basis with specified compounding. The phrase "6% compounded quarterly" means that the 6% is divided by 4 for the rate per interval (quarter) and there are 4 intervals per year. In mathematical symbolism

$$A = P\left(1 + \frac{0.06}{4}\right)^4$$

would be the balance for an amount P placed at 6% interest for 4 quarters or 1 year.

The stated problem asks for the accumulated balance on $1000 after 5 years at 6% interest compounded quarterly so the appropriate formula is

$$A = \$1000\left(1 + \frac{0.06}{4}\right)^{20} = \$1000(1.015)^{20} = \$1346.86$$

If the compounding is done monthly then the rate has to be divided by 12 and the number of compoundings increased to $12 \times 5 = 60$.

$$A = \$1000\left(1 + \frac{0.06}{12}\right)^{60} = \$1348.85$$

Example 9-13   In the previous problem what would be the balance at the end of 5 years if the compounding were increased to instantaneous compounding?

Solution: Start with the statement $A = P\left(1 + \dfrac{r}{k}\right)^{kt}$, where k represents the compounding rate (12 for monthly, 365 for daily) and kt is the number of compoundings over time.

Show me the money!

This looks so much like the definition of e, define k/r as n so the expression for the amount reads

$$A = P\left(1 + \frac{1}{n}\right)^{nrt}.$$

Now, knowing the laws of exponents, A can be written as $A = P\left[\left(1 + \dfrac{1}{n}\right)^{n}\right]^{rt}$.

As the number of intervals increases (k increases and k/r increases) the compunding approaches instantaneous and the expression in the brackets becomes

$$\lim_{n \to \infty}\left(1 + \frac{1}{n}\right)^{n} = e$$

So in the limiting case of instantaneous compounding the amount is $A = Pe^{rt}$.

For this problem with $P = \$1000$, $r = 0.06$, and $t = 5\,$years, the maximum balance for instantaneous compounding is

$$A = \$1000(e^{0.06 \cdot 5}) = \$1000(e^{0.3}) = \$1349.86$$

9-4   Logarithms

There are several definitions of logarithms. We will consider only the simpler ones. Further, we will consider only natural, or base e, logarithms.

The simplest definition of a logarithm is that it is a function that allows the exponential equation $y = e^x$ to be written in the form $x = \ldots$. The equation $y = e^x$ cannot be solved for x with conventional algebraic methods. The logarithmic function is the way out of this dilemma. The equivalence between exponents and logarithms is

$$y = e^x \quad \Leftrightarrow \quad \ln y = x$$

Although there are exponential equations other than base e, most of the exponential and logarithmic functions you encounter will be base e. Your hand calculator uses base e and base 10, though base 10 is used rarely. The notation ln, as opposed to log, specifies base e. The logarithmic equation just above is read as "log base e" or "ln e" or simply "log" with the later presuming that if the base were other than e it would be specified.

Run a few numbers on your calculator to become familiar with taking logarithms and calculating with exponents. This is not something you do every day and you don't want to make a calculating mistake on a test.

On your hand calculator raise e to a power, then take the ln of that number to return to the original power (number). [$e^3 = 20$; $\ln 20 = 3$] As you go through the problems in this chapter keep your calculator handy and practice "punching the numbers."

Example 9-14   Solve the equations $7 = e^x$ and $\ln u = 1.4$.

Solution: The equation $7 = e^x$ has to be switched to a logarithmic one: $x = \ln 7$ with $x = 1.95$.

The equation $\ln u = 1.4$ has to be switched to an exponential one: $u = e^{1.4}$ with $u = 4.06$.

Quick Tip

The key phrase to remember in switching from an exponential equation to a logarithmic equation and vice versa is "a logarithm is an exponent." The logarithm of something is a number and that number is the exponent of e.

There are manipulative laws for logarithms that parallel similar laws for exponents.

$$\ln(uv) = \ln u + \ln v$$

$$\ln\frac{u}{v} = \ln u - \ln v$$

$$\ln u^n = n \ln u$$

A little manipulation of exponents will verify the first law.

Set $k = \ln u$ and $l = \ln v$ so that $e^k = u$ and $e^l = v$. Form the product $uv = e^k \cdot e^l = e^{k+l}$, and convert to a logarithm equation $\ln uv = k + l = \ln u + \ln v$.

The derivative of $\ln x$ is

$$d(\ln x) = \frac{1}{x}\, dx \quad \text{or} \quad \frac{d}{dx}(\ln x) = \frac{1}{x}$$

Example 9-15 Find $\dfrac{dy}{dx}$ of $y = x^2 \ln x$.

Solution: Use the chain rule and go slowly.

$$dy = x^2 d(\ln x) + (\ln x)d(x^2) = x^2 \frac{1}{x}\, dx + (\ln x)(2x\, dx) = (x + 2x \ln x)dx$$

or

$$\frac{dy}{dx} = x(1 + 2 \ln x)$$

Example 9-16 Find $\dfrac{dy}{dx}$ for $y = \ln(2x^2 + 1)$.

Solution: $dy = \dfrac{1}{2x^2 + 1}\, d(2x^2 + 1) = \dfrac{1}{2x^2 + 1}(4x)dx = \dfrac{4x}{2x^2 + 1}\, dx$

or

$$\frac{dy}{dx} = \frac{4x}{2x^2 + 1}$$

Example 9-17   Find the derivative of $y = x^{1.5} \ln (x^2 + 2)$.

Solution: This looks bad. But, if you proceed slowly, applying the rules one at a time, the differentiation is not all that difficult. The hard part is proceeding logically. This is a product so write

$$dy = x^{1.5} d[\ln (x^2 + 2)] + \ln (x^2 + 2) dx^{1.5}$$

The differential of $\ln (x^2 + 2)$ is $\dfrac{1}{x^2 + 2} d(x^2 + 2) = \dfrac{2xdx}{x^2 + 2}$.

The differential of $x^{1.5}$ is $1.5x^{0.5} dx$. Putting it all together we write

$$dy = x^{1.5} \frac{2xdx}{x^2 + 2} + 1.5x^{0.5} \ln (x^2 + 2) dx \text{ or } \frac{dy}{dx} = \frac{2x^{2.5}}{x^2 + 2} + 1.5x^{0.5} \ln (x^2 + 2)$$

Example 9-18   Find the derivative of $y = e^{-x} \sin x$.

Solution: This is a product. Proceed methodically and the problem is not difficult.

$$dy = e^{-x} d(\sin x) + \sin x d(e^{-x}) = e^{-x}(\cos x) dx - e^{-x}(\sin x) dx$$

$$\frac{dy}{dx} = e^{-x}(\cos x - \sin x)$$

There is a simple rule for differentiating logarithmic functions that some authors use.

$$\text{If } y = \ln f(x) \text{ then } \frac{dy}{dx} = \frac{\frac{d}{dx}(f(x))}{f(x)}.$$

This is equivalent to using the chain derivative approach and the derivative of a logarithm as defined in the Mathematical Tables.

$$dy = \frac{1}{f(x)} d(f(x))$$

Verify for yourself that the two forms are equivalent by working Example 9-15 both ways.

Integration of the logarithmic function follows $\displaystyle\int \ln x \, dx = x \ln x - x$, which is used so rarely that we only give this one example.

Example 9-19   Find $y = \int \ln 2x \, dx$.

Solution: $y = \dfrac{1}{2} \int \ln 2x \, d(2x) = 2x \ln 2x - 2x$

9-5   Growth and Decay Problems

The growth and decay model is appropriate to many phenomena, such seemingly diverse problems as population growth, radioactive decay, the spread of disease, the cooling of a cup of coffee, and the number of yeast in a culture, just to name a few. Before working some problems, a very simple model of bacteria or yeast growth will be developed and worked through in detail. Notice the pattern in the problem. Many mathematical models of different phenomena parallel this one.

A simple experiment performed in elementary chemistry, biology, or physical science courses is the growth of bacteria or yeast. In this experiment a certain number (the number is often determined or measured by weight) of bacteria are placed in a nutrient environment. This means that the bacteria have optimum growing conditions, food, temperature, and the like. Their growth is then limited by their growth mechanism and not by anything external.

The bacteria grow by budding, one bacterium grows on another, splitting, each bacterium divides producing two identical bacteria, so that each bacterium over an average time period becomes two bacteria and these two repeat the same process in the same time period and on and on. At any time in the process the number of bacteria produced per unit of time is proportional to the number present. This is the mathematical statement of the growth model for bacteria. In symbolic form, dN/dt, the number produced per unit of time is proportional to the number present, kN.

$$\frac{dN}{dt} = kN$$

Solving this statement for N as a function of time is a calculus problem, and one we already have some experience with.

In practical terms the difficulty with the rate statement is that the N is on the wrong side of the equation. It needs to be associated with the dN if we are to make any progress toward a solution. A little algebra fixes this.

$$\frac{dN}{N} = kdt$$

Now integrate both sides of the equation. (Remember: $\int d(\ln x) = \int \frac{dx}{x}$ so $\int \frac{dx}{x} = \ln x$)

$$\int \frac{dN}{N} = \int kdt \quad \Leftrightarrow \quad \ln N = kt + C$$

The constant is required because there are no limits on the integrals.

In this problem, as with every problem in growth or decay, there is an initial amount of material. In this case there is an initial number of bacteria at the start of the experiment. Call this initial amount N_o. In the language of mathematics, at $t = 0$, $N = N_o$. Substitute these values into $\ln N = kt + C$.

$$\ln N_o = k(0) + C \quad \text{so} \quad C = \ln N_o$$

With the constant evaluated in terms of the initial amount of material the basic relation is

$$\ln N = kt + \ln N_o \quad \text{or} \quad \ln N - \ln N_o = kt \quad \text{or} \quad \ln \frac{N}{N_o} = kt$$

If you had any trouble manipulating the logarithms in the previous line, go back and review the manipulative rules for logarithms. At this point switch to an exponential format.

$$\ln \frac{N}{N_o} = kt \quad \Leftrightarrow \quad \frac{N}{N_o} = e^{kt} \quad \text{or} \quad N = N_o e^{kt}$$

This last statement correctly describes the model. The number of bacteria at any time starts out at N_o ($e^0 = 1$) and increases with time in an exponential manner.

This $N = N_o e^{kt}$ is the general growth law for something with growth proportional to the number present. Some text authors begin the discussion of growth and decay with this equation. This approach is simple but neglects the development of a mathematical model of a simple statement that "the growth of . . . is proportional to the number of . . . present at any time." A little reflection will convince you that this model fits many different phenomena.

Suppose in this bacteria growth problem that 100 bacteria are introduced into a growth environment (water, nutrients, and the like) and that 2 hr later the bacteria are separated from the environment or otherwise identified and that their number has increased to 130. Can this information be used to determine the growth law?

With these two numbers, N and N_o, and the time interval the constant k can be evaluated. The calculation is a little logarithm and exponent intense but follow along with your calculator. Substitute as follows:

$$130 = 100e^{2k} \quad \text{or} \quad 1.30 = e^{2k}$$

To solve for k switch $1.30 = e^{2k}$ to a logarithmic equation. (Remember: "A logarithm is an exponent.") The logarithm of something is an exponent so

$$\ln 1.30 = 2k \quad \text{or} \quad k = \frac{1}{2}\ln 1.30 = 0.13$$

The specific law governing the growth of these bacteria in this environment is

$$N = N_o e^{0.13t}$$

With this law it is possible to predict how many of these bacteria would be present, say, after 12 hr and starting with 50 bacteria. Put in the 50 for N_o and the 12 hr for t and we get

$$N = 50e^{0.13 \cdot 12} = 50(4.76) = 238$$

This model that starts with the statement that the growth rate is proportional to the amount present can, with a modest amount of calculus and initial information, be used to predict future growth.

There is a standard pattern to growth and decay problems that always works. The general procedure for these problems is outlined below.

1. Any problem where the number of events is proportional to the number of participants present can be written as dN/dt equals a constant ($+k$ for growth and $-k$ for decay) times the number: $\dfrac{dN}{dt} = \pm kN$.

2. Rearrange to $\dfrac{dN}{N} = \pm k\,dt$ and integrate $\displaystyle\int \dfrac{dN}{N} = \pm k \int dt$ to get $\ln N = \pm kt + N_0$. Take an initial number N_o, at $t = 0$, to evaluate $A = \ln N_o$, and write $\ln N = \pm kt + \ln N_o$.

3. Rearrange the equation to $\ln \dfrac{N}{N_o} = \pm kt$ and switch to an exponential format $\dfrac{N}{N_o} = e^{\pm kt}$ or $N = N_o e^{\pm kt}$.

4. One data point, a certain N at a specific time, allows calculation of k. (For example, a 20% increase in N_o in 1 hr means $1.2N_o = N_o e^{1k}$ or $1.2 = e^{1k}$. Switch to a logarithmic equation and $k = \ln 1.2 = 0.18$ and finally write $N = N_o e^{0.18t}$.)

5. With the calculation of k, the specific growth or decay equation is written for the same conditions that produced the initial data. With this specific growth or decay equation N at any time can be predicted.

Refer to this procedure in subsequent problems. It is a very logical procedure for growth and decay problems and it works. Growth and decay problems are favorite test problems. Know how to work them and especially know how to switch from exponential equations to logarithmic equations and vice versa and know how to take logs and perform exponentiation on your hand calculator.

Example 9-20   If "a fool and his money are soon parted," the rate at which it leaves is probably proportional to the amount remaining. If a certain fool starting with $20,000 starts gambling his money away and after 2 hr has lost $2000, how long will it take for him to loose 90% of the original amount?

Solution: The basic assumption in this problem is that the fool will loose in proportion to the amount he has at any time. Humans are a little harder to predict than bacteria, but this is a good assumption. Follow the procedural steps as written previously and be aware of the logic in the problem.

Step 1: The statement "the rate at which the fool looses money is proportional to the amount present" means that

$$\frac{dA}{dt} = -kA$$

Step 2: Rearrange, integrate, and evaluate the constant of integration with the initial data.

$$\frac{dA}{A} = -kdt, \quad \int \frac{dA}{A} = -k \int dt, \quad \ln A = -kt + C$$

At $t = 0$, the fool has $20,000, so $\ln 20{,}000 = -k(0) + C$ and $C = \ln 20{,}000$.

Now the equation reads

$$\ln A = -kt + \ln 20{,}000$$

Step 3: Rearrange and switch to exponents.

$\ln \dfrac{A}{20,000} = -kt$, and switching to exponents $\dfrac{A}{20,000} = e^{-kt}$ or $A = 20,000e^{-kt}$.

Step 4: Use the given data to determine k.
At $t = 2\,\text{hrs}$, A has declined to 18,000, so put these numbers into the amount statement and find k.

$$18,000 = 2000e^{-2k}, 0.9 = e^{-2k}$$

Switch to logarithms to solve this equation for k.

$$-2k = \ln(0.9), k = -\dfrac{\ln(0.9)}{2} = 0.053$$

As you were following along this problem and "punching the numbers," so you would be very proficient at this logarithm and exponent calculating for the test on this topic, you may have noticed that your calculator displayed a negative number for ln(0.9). This is correct. In the original statement of the problem, $dA/dt = -kA$ so that the calculation of k should produce a positive number. The reason for the ln of numbers less than 1 being negative has to do with one of those other definitions of the ln and will be taken up shortly.

Step 5: The specific equation for this situation is $A = 20,000e^{-0.053t}$.

The time for 10% remaining is the time for A to reach 2000. Substitute for $A = 2000$ and solve for t.

$$2000 = 20,000e^{-0.53t}, 0.1 = e^{-0.053t}$$

Switching to logarithms, $\ln(0.1) = -0.053t$ or $t = -\dfrac{1}{0.053}\ln(0.1) = 43\ \text{hrs}$.

Based on this model, it would take this particular fool 43 hr to loose 90% of an original amount of $20,000.

Example 9-21   Hot or cold objects cool down or heat up to the temperature of their surroundings. The temperature difference, ΔT, between the object and its surroundings decreases over time in proportion to that temperature difference. This is Newton's law of cooling. If a cup of coffee cools from 85°C to 80°C in 2 minutes in room temperature surroundings, how long does it take for the coffee to cool from 85°C to 30°C (20°C is room temperature)?

Solution: Don't be fooled by the wording of this problem. It is not the temperature that is important, but the difference in temperature between the coffee and its surroundings. The definitive statement is "the change in the temperature difference is proportional to the difference." Call ΔT the difference in temperature between the coffee and its surroundings. The mathematical statement of Newton's law of cooling then is

$$\frac{d(\Delta T)}{dt} = -k(\Delta T)$$

Rearrange and integrate.

$$\frac{d(\Delta T)}{\Delta T} = -kdt, \qquad \int \frac{1}{\Delta T} d(\Delta T) = -k \int dt, \ln(\Delta T) = -kt + C$$

When the coffee starts cooling, the temperature difference is 65°C = (85 − 20)°C so $\ln 65 = -k(0) + C$ and $C = \ln 65$ so the equation becomes

$$\ln(\Delta T) = -kt + \ln 65 \quad \text{or} \quad \ln \frac{\Delta T}{65} = -kt$$

Switching to exponents we write

$$\frac{\Delta T}{65} = e^{-kt} \quad \text{or} \quad \Delta T = 65e^{-kt}$$

Be careful with this next step. The temperature changes by 5°C so the temperature difference is now 60°C, and this occurs over 2 min so put in these values and evaluate k.

$$60 = 65e^{-2k}, \quad \frac{60}{65} = e^{-2k}, \quad \text{and switching to logarithms}$$

$$-2k = \ln \frac{60}{65}, \quad k = -\frac{1}{2} \ln \frac{60}{65} = 0.04$$

The temperature difference statement is now

$$\Delta T = 65e^{-0.04t}$$

Be careful again. Go back and read the question and make sure you understand that the desired time is for the temperature to reach a 10°C difference between the coffee and its surroundings. Use the 10°C temperature difference and calculate the time.

$$10 = 65e^{-0.04t}, \frac{10}{65} = e^{-0.04t}, \qquad -0.04t = \ln\frac{10}{65}, t = -\frac{1}{0.04}\ln\frac{10}{65} = 47 \text{ minutes}$$

It takes 47 min for this cup of coffee to cool to within 10° C of room temperature.

Example 9-22 A wildlife manager needs to reach a 10,000 population of mule-eared deer in a certain habitat in 6 years. There are presently no deer in the habitat. The environment is such that the deer can grow without being limited by their environment. This means that the growth of the deer population will be proportional to the population, $dP/dt = kP$. In order to determine the growth equation, 100 deer are introduced into the habitat. There are half males and half females, the same ratio as when they reproduce. At the end of the year there are 130 deer. How many deer need to be introduced to the habitat to achieve the 10,000 goal in the remaining 5 years?

Solution: The first part of the problem is to determine the growth equation. Starting with

$$\frac{dP}{dt} = kP, \text{ the general growth equation is } P = P_o e^{kt}$$

The 100 deer population grows to 130 in 1 year so put this data into $P = P_o e^{kt}$ and determine k.

$$130 = 100e^{1k} \text{ or } 1.30 = e^{1k} \text{ so that on switching, } \ln 1.30 = k \text{ or } k = 0.26$$

The specific growth law for these deer in this habitat is $P = P_o e^{0.26t}$. Now solve for the initial number P_o needed to produce the 10,000 population in 5 more years.

$$10,000 = P_o e^{0.26 \cdot 5} = P_o e^{1.30} = 3.67 P_o \text{ or } P_o = 2725$$

This number minus the 130 already there, or $2725 - 130 = 2595$, deer need to be introduced to achieve the 10,000 goal in the prescribed time.

9-6   The Natural Logarithm

Another and more formal definition of the natural logarithm relates the $\ln x$ to portions of the area under the curve $y = 1/t$.

The natural logarithm of any number x is defined by $\ln x = \int_1^x \dfrac{dt}{t}$.

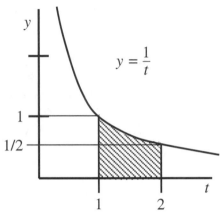

Fig. 9-4

The curve $y = 1/t$ and the graphical depiction of $\ln x$ as the area under the curve are shown in Fig. 9-4. The area under the curve between $x = 1$ and $x = 2$ is $\ln 2$. This area can be determined by taking as many narrow trapezoids or rectangles approximating this area as necessary to achieve a desired precision.

Look more closely at the piece of the curve between $t = 1$ and $t = 1.1$ (Fig. 9-5). The area under this part of the curve is approximated by the area of the rectangle $0.10 \times 0.91 = 0.091$ and the (area of the) small triangle $(1/2)(0.10) \times (0.09) = 0.0045$.

The total area of this rectangle and triangle is 0.0955, thus $\ln 1.1 = 0.0955$. The logarithm produced in most hand calculators is 0.0953, just a little bit smaller than this number as is expected from the shape of the curve.

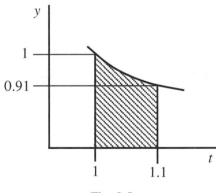

Fig. 9-5

For numbers less than 1, the integration in dx is in the negative direction. This produces the negative numbers for logarithms of numbers less than 1.

The function $y = \ln x$ is shown in Fig. 9-6. Referring to Fig. 9-4 and remembering that the definition of $\ln x$ is the area under the curve, note the following features of the $\ln x$ curve.

- Only positive values of x are allowed.

- $\ln 1$ is zero (no area).

- As x goes from 1 to 0, $\ln x$ (the area) goes from 0 to large negative numbers.

- As x goes from 1 to large positive numbers, $\ln x$ increases with the increase less and less as x goes to large positive numbers.

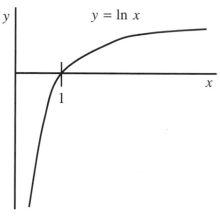

Fig. 9-6

The connection between this definition of the natural logarithm and the constant e is amazing! The constant e raised to the power equal to the area under the curve is equal to the upper limit of the integral.

The precise calculation of the area corresponding to an upper limit for the integral of 1.1 is 0.0953.

$$e^{area} = \text{upper limit of area calculation}$$

$$\ln(\text{upper limit of area calculation}) = \text{area}$$

Verify for yourself that $e^{0.0953} = 1.1$, and that $\ln 1.1 = 0.00953$.

Again, a reasonably simple area problem in calculus produces a number that occurs other places in nature.

9-7   Limited Exponential Growth

In many real-life problems growth is limited. Exponential models are used to describe limited growth. The simplest model for limited growth involving exponentials is one in the form $N = N_o(1 - e^{-kt})$. This statement is the result of a rate equation that is more complicated than the ones discussed earlier. This form of limited rate equation is shown in Fig. 9-7. Note that the N starts at zero at time zero. In $N = N_o(1 - e^{-kt})$ when $t = 0$, $e^0 = 1$ and $N = N_o(1 - 1) = 0$. After a long time, e^{-kt} or $1/e^{kt}$ becomes very small so that N approaches N_o. The slope of the curve is the derivative, or

$$dN = d(-N_o e^{-kt}) = -N_o d(e^{-kt}) = -N_o e^{-kt} d(-kt) = -N_o(-k)e^{-kt}dt$$

$$\frac{dN}{dt} = N_o k e^{-kt} = N_o k \frac{1}{e^{kt}}$$

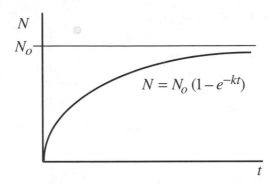

Fig. 9-7

At $t = 0$ the slope is (positive) $N_o k$ and as time goes on the slope decreases. This type of curve is sometimes called the learning curve because it describes someone learning a skill and eventually reaching a limit in productivity with that skill.

Example 9-23　Workers hired to assemble sewing machines become more skilled with experience. The most experienced workers can assemble 10 sewing machines per day. The learning curves are different for different workers but they all eventually reach a peak production of 10 sewing machines per day. A newly hired assembler learns to assemble 5 sewing machines per day after 6 working days. How long will it take for this worker to reach 9 sewing machines per day?

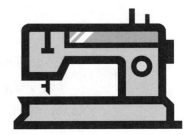

Solution:　The simple learning curve model is most appropriate for this problem. $N_o = 10$ is the maximum rate of sewing machine assembly. The general equation governing the number of sewing machines assembled per day then is

$$N = 10(1 - e^{-kt}) = 10 - 10e^{-kt}$$

The k can be determined with the information that after 6 days this particular worker can assemble 5 sewing machines per day. Substitute $N = 5$ and $t = 6$ and solve for k.

$5 = 10 - 10e^{-6k}$, $-5 = -10e^{-6k}$, $e^{-6k} = \dfrac{1}{2}$, and switching to logarithms

$$\ln\frac{1}{2} = -6k, \quad k = -\frac{1}{6}\ln\frac{1}{2} = 0.115$$

The specific learning equation for this worker is

$$N = 10(1 - e^{-0.115t})$$

The time for this worker to achieve a rate of 9 sewing machines per day is obtained by putting in 9 for N and solving for t.

$$9 = 10 - 10e^{-0.115t}, \quad -1 = -10e^{-0.115t}, \quad 0.1 = e^{-0.115t}$$

and switching to logarithms

$$-0.115t = \ln 0.1, \quad t = -\frac{1}{0.115}\ln 0.1 = 20 \text{ days}$$

It will take this worker 20 days to be able to assemble 9 sewing machines per day.

This type of limited growth relationship is most often encountered in the analysis of electric circuits. The simplest circuit to consider is one containing a battery, a resistor and a capacitor. The battery is the source of electric current. The resistor opposes the flow of that current. And the capacitor stores charge and as it stores this charge assumes a voltage ($q = CV$). A schematic diagram of a typical circuit is shown in Fig. 9-8.

Charging

Assume the capacitor, C, has zero charge and voltage, and place the switch, S, in the charging (up) position. When the battery voltage is applied to the R and C, voltages across these components vary with time. Initially there is a large

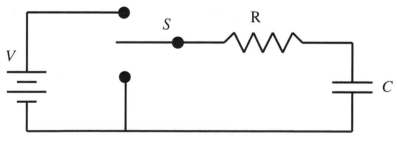

Fig. 9-8

current transporting charge to the capacitor. As charge builds up on the capacitor a voltage develops across the capacitor, and the current decreases. The parameters V, R, C, and the charge, q, and current, $i = dq/dt$ control the voltages, current, and charge within the circuit.

The analysis starts with a voltage statement that is true for any instant of time.

$$V = iR + \frac{q}{C} \quad \text{or} \quad CV - q = RC\frac{dq}{dt} \quad \text{or} \quad \frac{dq}{q - CV} = -\frac{1}{RC}dt$$

In this statement, remember that $i = dq/dt$. The last equation can be integrated using a change of variable. Replace $q - CV$ with x so $dq = dx$.

Then $\int \frac{dx}{x} = -\frac{1}{RC}\int dt$ and the integral of the left side is $\ln(x) = \ln(q - CV)$ so

$$\ln(q - CV) = -\frac{1}{RC}t + \ln K_1$$

Notice that the logarithm equation goes into an exponential equation. This is not an every day mathematical operation. Review the previous chapters on exponential to logarithm transforms if you need to brush up on this procedure. Rewrite the equation

$$q = CV + K_1 e^{-t/RC}$$

and apply the condition that at $t = 0$, $q = 0$, or $0 = CV + K_1$ or $K_1 = -CV$ so finally

$$q = CV(1 - e^{-t/RC})$$

Obtaining this expression for q looks easy and it is if you remember to choose the constant as a logarithm, thus making the equations easier to manipulate, remember how to switch from a logarithmic equation to an exponential equation, and apply the initial conditions correctly.

The voltage across the capacitor is related to the value of C and the charge on the capacitor at any time according to $q = CV$. The voltage across the capacitor in this circuit is

$$\frac{q}{C} = V_C = V(1 - e^{-t/RC})$$

The current in the resistor is

$$i = \frac{dq}{dt} = \frac{V}{R}e^{-t/RC}$$

The voltage across the resistor is

$$V_R = iR = Ve^{-t/RC}$$

There are two graphs that are helpful in understanding the situation: one is q vs. t, and the other is i vs. t. These are shown in Fig. 9-9.

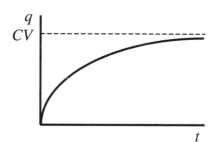
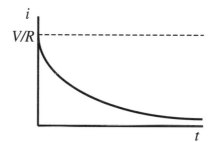

Fig. 9-9

Notice in the curve of q v.s t, that at $t = 0$, $q = 0$, and as $t \to \infty$, $q \to CV$.

In the curve of i vs. t, at $t = 0$, $i = V/R$, and as $t \to \infty$, $i \to 0$.

At $t = 0$ the charge on the capacitor is zero and the current is maximum and as t goes to infinity the current goes to zero and the charge reaches its maximum of CV.

The Time Constant

In biological systems that grow exponentially the systems are often characterized by giving the doubling time, the time for the system to double in number, size, or mass. In electrical systems that grow exponentially the systems are characterized by a time constant, the time to make the exponent of e equal to 1. The time constant for this circuit is RC.

Look at the equation above and the q vs. t graph. When $t = RC$, $1 - e^{-1} = 0.63$ and the charge on the capacitor has risen to 0.63 of its final value. A similar statement can be made about the voltage on the capacitor. After one time constant, the voltage on the capacitor is 0.63 of the battery voltage.

The current, meanwhile, has in one time constant dropped to $e^{-1} = 0.37$ of its initial value.

Example 9-24 A 10 kΩ resistor and a 20 μF capacitor are placed in series with a 12 V battery as shown in Fig. 9-8. Find the charge on the capacitor, the current, and the voltages on the capacitor and resistor at the instant the switch is closed, $t = 0$.

Solution: At $t = 0$ the charge on the capacitor is zero.

At $t = 0$, the current is $i = V/R = 12\,V/10 \times 10^3\,\Omega = 1.2 \times 10^{-3}\,A$.

At $t = 0$, the voltage on the capacitor is zero (it has no charge) and the entire battery voltage of 12 V is across the resistor.

Example 9-25 For the circuit of the previous problem, find the time constant, and the charge, current, and V_R, and V_C at a time equal to one time constant.

Solution: The time constant is $RC = 10 \times 10^3\,\Omega \cdot 20 \times 10^{-6}F = 0.20\,s$. The charge on the capacitor at $t = 0.20\,s$ is

$$q|_{t=RC} = CV(1 - e^{-1}) = 12\ V(20 \times 10^{-6}F)0.63 = 1.5 \times 10^{-4}C$$

The current at $t = 0.20s$ is

$$i|_{t=RC} = \frac{V}{R}e^{-1} = \frac{12\ V}{1.0 \times 10^4\,\Omega}\,0.37 = 4.4 \times 10^{-4}A$$

The voltage across the capacitor is $V_C = V(1 - e^{-1}) = 12\ V \cdot 0.63 = 7.6\ V$. The voltage across the resistor is $12\,V - 7.6\,V = 4.4\,V$.

These problems can be deceptively easy. Be sure you know how to manipulate the exponents on your calculator. Don't get a test problem wrong because you did not practice all the steps in the problem and were unfamiliar with manipulating exponents on your calculator.

Example 9-26 For the circuit of the previous two problems, how long does it take for the capacitor to reach 80% of its final charge?

Solution: This problem is similar to radioactive decay problems where we want to know the time for half the material to decay. There is a fair amount of algebraic manipulation that is easier to follow without numbers, so the problem will be worked as far as possible with symbols.

Start with $q = CV(1 - e^{-t/RC})$ and note that the final (fully charged) q is equal to CV. Mathematically, when $t \to \infty$, $e^{-t/RC} \to 0$ and $1 - e^{-t/RC} \to 1$ so $q \to CV$.

To find the time for 80% charge set q equal to 80% of the final charge, or $q = 0.80CV$, and solve for t.

$$0.80CV = CV(1 - e^{-t/RC}) \quad \text{or} \quad 080 = 1 - e^{-t/RC} \quad \text{or} \quad e^{-t/RC} = 0.20$$

For convenience switch to positive exponents so

$$1/e^{t/RC} = 0.20 \quad \text{or} \quad e^{t/RC} = 1/0.20 = 5$$

In order to solve for t, switch the exponential equation to a logarithmic equation. One of the functions of logarithms is to solve for variables in exponents.

$$t/RC = \ln 5 \quad \text{or} \quad t = RC \ln 5$$

Now put in the values for R and C.

$$t = RC \ln 5 = 1.0 \times 10^4 \,\Omega \cdot 20 \times 10^{-6} \text{F} \cdot \ln 5 = 0.32 \text{ s}.$$

As a check note that $1 - e^{-t/RC} = 1 - e^{-0.32/0.20} = 0.80$

Discharging

After the capacitor is left to charge for a long time (many time constants) the charge is CV. Move the switch to the discharge position (down in Fig. 9-8) where R and C are in series. When the charged capacitor and resistor are placed in series the charged capacitor acts as a battery. The voltage on the capacitor is q/C, and this voltage appears across the resistor as iR. As time goes on, the charge on the capacitor is depleted and the current drops (eventually) to zero. The charge decays according to

$$q = CVe^{-t/RC}$$

and the current according to

$$i = \frac{dq}{dt} = \frac{d}{dt} CVe^{-t/RC} = -\frac{V}{R} e^{-t/RC}$$

The negative sign indicates that the current in the resistor is in the opposite direction from the charging situation, which must be the case.

Mathematical analysis of the discharge circuit starts with a voltage type of statement similar to the one used in the charging case.

$$iR + \frac{q}{C} = 0 \quad \text{or} \quad \frac{dq}{dt} = -\frac{1}{RC}q \quad \text{or} \quad \frac{dq}{q} = -\frac{1}{RC}dt$$

and solving

$$\ln q = (-1/RC)t + \ln K_2$$

Again notice that the choice of constant is very convenient because

$$\ln\frac{q}{K_2} = -\frac{1}{RC}t \quad \text{or} \quad \frac{q}{K_2} = e^{-t/RC} \quad \text{or} \quad q = K_2 e^{-t/RC}$$

Now impose the initial condition. At $t = 0$, $q = CV$, so $K_2 = CV$ and

$$q = CVe^{-t/RC}$$

The charge on the capacitor decays with the same time constant, RC.

Example 9-27   If the circuit used in the previous problems is placed in the discharge mode, how long does it take for the circuit to discharge to 50% of its original (total) charge?

Solution: Solve $q = CVe^{-t/RC}$ for t when $q = 0.50\,CV$.

$$0.50\,CV = CVe^{-t/RC} \quad \text{or} \quad 0.50 = e^{-t/RC}$$

It is more convenient to write 0.50 as 1/2, so when the statement is converted to logarithms,

$$\ln 0.50 = -\frac{t}{RC} \quad \text{is} \quad \ln 1 - \ln 2 = -\frac{t}{RC} \quad \text{and since } \ln 1 = 0$$

$$t = RC \ln 2 = 1.0 \times 10^4\,\Omega \cdot 20 \times 10^{-6}\,\text{F} \cdot \ln 2 = (0.20)(0.69\,\text{s}) = 0.14\,\text{s}$$

In terms of time constants this would be $\ln 2$, $(RC)\ln 2$, time constants, or 0.69 of a time constant.

It makes sense that our answer is less than one time constant since it takes less time for the charge to decline to 50% of its initial value than to 37% ($1/e$) of its initial value.

Example 9-28 An RC circuit is observed during discharge to have an initial capacitor potential of 100 V and after 3.0 s to have a potential of 20 V. How long will it take for the capacitor to discharge to 1.0 V.

Solution: The voltage across the capacitor at any time is determined by $q = CVe^{-t/RC}$ rewritten as q/C, or $V_C = V_o e^{-t/RC}$, where V_o is the voltage at $t = 0$.

Take $V_o = 100$, $V_C = 20$ V, and $t = 3.0$ s and write

$$20\,\text{V} = 100\,Ve^{-3.0/RC} \quad \text{or} \quad 2/10 = e^{-3.0/RC} \quad \text{or} \quad e^{3.0/RC} = 5$$

Switching to logarithms

$$\frac{3.0}{RC} = \ln 5 \quad \text{or} \quad \frac{3.0}{\ln 5} = RC \quad \text{or} \quad RC = 1.9\,\text{s}$$

Knowing this number, the specific decay law for this circuit can be written

$$V = (100\ \text{V})e^{-t/1.9}$$

Now calculate the time for the voltage to drop to 1.0 V.

$$1.0\,\text{V} = 100\,Ve^{-t/1.9} \quad \text{or} \quad (1/100) = e^{-t/1.9}$$

Switching to logarithms

$$-\ln 100 = -(t/1.9) \quad \text{or} \quad t = 1.9 \cdot \ln 100 = 8.6\ \text{s}$$

Go back over this problem and note the procedure.

1. After reading the problem, the general law (equation) was written down, $V_C = V_o e^{-t/RC}$.
2. Next the data from the problem (100 V going to 20 V in 3.0 s) was used to find RC.
3. With RC the specific law for this problem was written; $V = 100\,Ve^{-t/1.9}$.
4. Finally, with this specific law the predictive calculation was performed to find the time for the 100 V to decay to 1.0 V.

This analysis procedure is typical of growth and decay problems in general. Be familiar with the steps in this procedure. It will keep you from getting lost and not knowing how to proceed in problems like this.

9-8   The Logistic Function

Another type of exponential function used to describe limited growth has the form

$$R = \frac{B}{1 + Ae^{-kt}}$$

At $t = 0$, $R = \dfrac{B}{1 + A}$. This is the present rate or number, whatever R represents.

As t goes to infinity, $e^{-kt} = 1/e^{kt}$ goes to 0 and R approaches B. B is the maximum rate or number.

This curve has the general shape shown in Fig. 9-10.

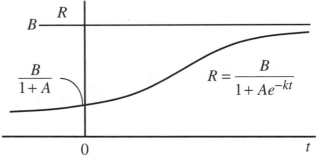

Fig. 9-10

Many industries follow this type of a growth curve. When a new product is introduced there is considerable demand, but as more and more people acquire, the product sales drop to a level determined by the number of new people entering the marketplace and replacement of old or outdated product. The automobile industry is an excellent example of this type of growth.

The exponential part of the definition is sometimes written as b^x. This form is workable if data defining the function is available for small values of x. Look at the logistic function in this form and see how it is convenient for evaluating the constants for small x.

$$f(x) = \frac{N}{1 + Ab^{-x}}$$

Multiply the right side of this function up and down by b^x.

$$f(x) = \frac{Nb^x}{(1 + Ab^{-x})b^x} = \frac{N}{b^x + A} b^x$$

For small values of x
the term b^x is approximately 1, so the function can be written as

$$f(x) \approx \frac{N}{1 + A} b^x \quad \text{for small } x$$

This use of data for small x also works using the e^{kt} form. The exponential is the most popular form for the logistic equation.

Example 9-29 A microchip production line has a theoretical maximum output of 400 chips per day. The factory production managers know from experience that new microchip production lines reach maximum production according to

$$R = \frac{400}{1 + 3e^{-0.08t}}$$

where R is in hundreds of chips per day.

Sketch the function and find the production rate on the first day of operation, the tenth day of operation, and finally, the maximum rate of production.

Solution: The curve is the standard one shown in Fig. 9-11.

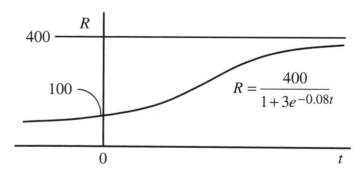

Fig. 9-11

Take $t = 0$ for the first day of production so we have

$$R = \frac{400}{1 + 3(1)} = \frac{400}{4} = 100 \text{ chips per day}$$

Take $t = 10$ for the tenth day of production so

$$R = \frac{400}{1 + 3e^{-0.08(10)}} = \frac{400}{1 + 3(0.45)} = 170 \text{ chips per day}$$

As t goes to infinity, the denominator in the rate equation goes to 1 and the maximum rate goes to 400 chips per day.

Example 9-30   The fish population on a fish farm are assumed to grow according to a logistic curve. The farm will support a maximum of 2000 fish. The farm is started with 200 fish ($t = 0$) and at the end of 2 months ($t = 2$) there are 350 fish. What is the fish population at 6 months?

Solution: Since there is data for small t look at both forms of the logistic function; the basic function and the approximation for small t.

$$F(t) = \frac{N}{1 + Ae^{-kt}} \quad \text{or} \quad F(t) \approx \frac{N}{1 + A} e^{kt} \text{ for small } t$$

The maximum number of fish is 2000, so $N = 2000$.

At $t = 0$ there are 200 fish so

$$200 = \frac{2000}{1 + A} \quad \Rightarrow \quad A = 9$$

Now use the condition that at $t = 2$, there are 350 fish.

$$350 = \frac{2000}{1 + 9e^{-2k}} \quad \Rightarrow \quad 1 + 9e^{-2k} = \frac{2000}{350}$$

$$9e^{-2k} = 4.71 \quad \Rightarrow \quad e^{-2k} = 0.524$$

With a little more algebra and switching to logarithms.

$$e^{2k} = 1.91 \quad \Rightarrow \quad \ln 1.91 = 2k \quad \Rightarrow \quad k = 0.32$$

The function can now be written

$$F(t) = \frac{2000}{1 + 9e^{-0.32t}}$$

and evaluated at the end of 6 months.

$$F(6) = \frac{2000}{1 + 9e^{-1.94}} = 872 \text{ fish}$$

It's a Wrap

✔ Interval and instantaneous compound interest

✔ Know how to differentiate logarithms and exponents

✔ Know how to integrate logarithms and exponents

✔ Differentiate and integrate trigonometric functions

✔ Solve growth and decay problems

✔ Use spread sheets to find rates and predict total returns

✔ Solve limited growth exponentials

✔ Solve problems that fit a logistic model

Test Yourself

PROBLEMS

1. Differentiate $f(x) = \ln(x - 1)$, $f(x) = x \ln x$, and $f(x) = (x^2 + 1) \ln x$.

2. Integrate $\int \frac{x\, dx}{x^2 + 1}$, $\int \left(3e^x + \frac{2}{x} - \frac{x^2}{2}\right)dx$ and $\int x^3 e^{x^4 + 2} dx$.

3. An amount of $2000 is invested at 12% with compounding quarterly. Determine how long it takes to grow the $2000 to $5000.

4. The current and voltage in an *R-L* (resistance and inductor) circuit behave in much the same way as for an *R-C* circuit. The voltage statement valid at every instant after the switch is closed is $V - iR - L\frac{di}{dt} = 0$.

    Solve this equation for i and di/dt for a 60 V battery connected to a 50×10^{-3} H inductor and a 180 Ω resistor.

5. What is the time constant for the circuit of problem 4?

6. Radioactive materials decay in direct proportion to the amount of material present. Radioactive Polonium decays with a half-life of 140 days. How much of a 300 g sample remains after 360 days. You may want to review the procedure for growth and decay problems in the text before working this problem.

7. For the previous problem how long does it take for 90% of the material to decay?

8. $100,000 invested in an oil well produces the following results for the first 6 months: $R(1) = 2700$, $R(2) = 2150$, $R(3) = 2010$, $R(4) = 1800$, $R(5) = 1760$,

and $R(6) = 1680$. Find the equation that fits this data and find the rate of decline at $R(5)$.

9. For the previous problem, what is the expected return over the first 12 months?

10. The number of pet strollers sold by a certain company in the 1995 to 2003 range was $n(t) = 0.04t^2 + 0.3t + 15$ with t measured in thousands of strollers. The year 1995 corresponds to $t = 5$ and so on up to $t = 13$ for 2003. What was the rate of stroller sales in the year 2000?

11. Use the logistic formula with base e to describe the growth of tattoo parlors as the industry matures. Assume that a 3 million population metropolis can support 200 parlors. In 1990, $t = 0$, there are 6 parlors. In 1993, $t = 3$, there are 12 parlors. Determine the specific logistic formula and predict the number of parlors in 1996, $t = 6$.

12. Sales of the Supertimewaster video game are following the logistic model given below, where q is the number of games sold per month and t is the time in months. What is the rate of sales at 10 months?

$$q(t) = \frac{5000}{1 + 2e^{-0.4t}}$$

ANSWERS

1. $df(x) = \dfrac{1}{x - 1} d(x - 1) \quad \Rightarrow \quad \dfrac{df}{dx} = \dfrac{1}{x - 1}$

   $df(x) = xd \ln x + \ln x dx \quad \Rightarrow \quad \dfrac{df}{dx} = 1 + \ln x$

   $df(x) = (x^2 + 1)\dfrac{dx}{x} + \ln x (2xdx) \quad \Rightarrow \quad \dfrac{df}{dx} = \dfrac{x^2 + 1}{x} + 2x \ln x.$

2. Let $u = x^2 + 1$ so $du = 2xdx$ $\displaystyle\int \dfrac{xdx}{x^2 + 1} \Rightarrow \dfrac{1}{2}\int \dfrac{du}{u} = \dfrac{1}{2}\ln u = \dfrac{1}{2}\ln |x^2 + 1|$

   $3e^x + 2\ln |x| - \dfrac{x^3}{6}$

   $\displaystyle\int x^3 e^{x^4 + 2} dx$ can be integrated with a change of variable. First rewrite the

   integral as $\dfrac{1}{4}\displaystyle\int 4x^3 e^{x^4 + 2} dx$. With a new variable $u = x^4 + 2$ and $du = 4x^3 dx$

   the integral is now $\dfrac{1}{4}\displaystyle\int e^u du = e^u$ so $\displaystyle\int x^3 e^{x^4 + 4} dx = \dfrac{1}{4}[e^{x^4 + 2}].$

3. Start with the statement $A = P\left(1 + \dfrac{r}{k}\right)^{kt}$, where P is the initial amount, r is the annual rate, k is the compounding interval and t is the number of years. Enter the numbers and the expression looks like $A = (\$2000)\left(1 + \dfrac{0.12}{4}\right)^{4t}.$

Now impose the requirement for A to reach \$5000: $\$5000 = (\$2000)(1.03)^{4t}$. With a little manipulation, $2.5 = (1.03)^{4t}$. This is one of those inocuous looking equations that can drive you crazy! If the base were e, it would be appropriate to take logarithms. Since the base is 1.03, we need either base 1.03 logarithms, a mechanism for switching bases of logarithms, or another method for solving for t. It is possible to switch bases of logarithms but in this case a few key strokes on a calculator or a spread sheet will solve the problem nicely. Raise 1.03 to a few powers and find what power produces 2.5 to the two significant figures given in the problem. About 10 key strokes and a fraction of a minute produces a number of 31.1 for the exponent $4t$ so about 7.8 years are necessary at these conditions to grow \$2000 to \$5000.

4. The solution is similar to the one for the R-C circuit done in the text. A review of that more detailed analysis may be helpful before continuing.

Rewrite this equation as $\dfrac{di}{dt} = \dfrac{V}{L} - i\dfrac{R}{L}$ or $\dfrac{di}{dt} = \left(\dfrac{V}{R} - i\right)\dfrac{R}{L}$.

Switching to a more convenient form (separate the variables) for integration

$$\int \frac{di}{i - (V/R)} = -\int \frac{R}{L}\, dt$$

Integrating with a change of variable to $i - V/R$, $\ln\left(i - \dfrac{V}{R}\right) =$

$-\dfrac{R}{L}t + \ln K$ or $\ln(i - V/R) - \ln K = -\dfrac{R}{L}t$ or $\ln\left[\dfrac{i - V/R}{K} = -\dfrac{R}{L}t\right]$

Switching to exponentials $i - (V/R) = Ke^{-Rt/L}$ or $i = (V/R) + Ke^{-Rt/L}$
The constant, K, can be evaluated in either of these equations by imposing the initial condition that at $t = 0$, $i = 0$, $0 = V/R + K$ so $K = -V/R$ and $i = \dfrac{V}{R}(1 - e^{-Rt/L})$ and $\dfrac{di}{dt} = \dfrac{V}{L}e^{-Rt/L}$ Now put in the numbers.

First find $\dfrac{R}{L} = \dfrac{180\,\Omega}{50 \times 10^{-3}\,H} = 3.6 \times 10^3\,(1/s)$

$i = \dfrac{60\,V}{180\,\Omega} e^{-3.6\times10^3 t}$

$\dfrac{di}{dt} = \dfrac{60\,V}{50 \times 10^{-3}\,H} e^{-3.6\times10^3 t}$.

5. The time constant $\tau = \dfrac{L}{R} = \dfrac{50 \times 10^{-3}\,H}{180\,\Omega} = 2.8 \times 10^{-4}\,s$.

6. The basic equation is $\dfrac{dN}{dt} = -kN$. The minus sign is for decay. Rewrite and integrate. $\dfrac{dN}{N} = -kdt$, $\ln N = -kt + \ln K$. At $t = 0$, $N = 300$ so $K = 300$.

$\ln\dfrac{N}{300} = -kt \implies N = 300e^{-kt}$ The amount goes to 150 g in 140 days so

$$150 = 300e^{-140k} \quad \Rightarrow \quad \frac{1}{2} = e^{-140k} \quad \Rightarrow \quad -\ln 2 = -140k \quad \Rightarrow \quad k = 0.005$$

This gives the specific relationship $N = 300e^{-0.005t}$ and for 360 days $N = 300e^{-1.8} = 50\,g$

7. It is easiest to go back to the basic equation and write $0.10N_o = N_oe^{-0.005t}$. So $\ln 0.10 = -0.005t \quad \Rightarrow \quad t = 460$ days.

8. Place the data on a spread sheet, then have the spread sheet draw the graph and write the equation that best fits the data.

    The equation for cash return is $R = 49t^2 - 529t + 3120$ and the rate of return at $R(5)$ is the differential of this equation evaluated at 5.

$$\frac{dR}{dt} = 98t - 529 \qquad \frac{dR}{dt}\Big|_{t=5} = 98(5) - 529 = -39 \text{ dollars per month}$$

R	t
1	2700
2	2150
3	2010
4	1800
5	1760
6	1680

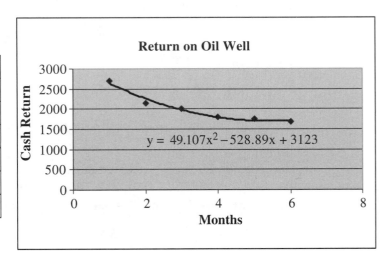

Return on Oil Well

$y = 49.107x^2 - 528.89x + 3123$

9. Based on the data and the rate equation the estimated return for the first 12 months is the integral of the rate statement from 0 to 12.

$$\int_0^{12} (49t^2 - 529t + 3123)dt = \left[\frac{49t^3}{3} - \frac{529t^2}{2} + 3123t \right]\Big|_0^{12}$$

$$= \left[\frac{49(12)^3}{3} - \frac{529(12)^2}{2} + 3123(12) \right] = 28368 \text{ dollars}$$

10. The year 2000 corresponds to $t = 10$ so take the derivative of n and evaluate at $t = 10$ $\frac{dn}{dt} = 0.08\,t + 0.3 \quad \Rightarrow \quad \frac{dn}{dt}\Big|_{10} = 0.8 + 0.3 = 1.1$ or 1100 strollers

11. Start with the logistic formula $f(t) = \dfrac{N}{1 + Ae^{-kt}}$ and multiply up and down by e^{kt} so $f(t) = \dfrac{N}{e^{kt} + A}e^{kt}$. $N = 200$, the estimated maximum number of parlors. Substitute $t = 0$ and $f(t) = 6$ to evaluate A. $6 = \dfrac{200}{1 + A}$ for $A = 32$

The logistic formula now reads $f(t) = \dfrac{200}{1 + 32e^{-kt}}$ and substituting

$t = 3$ and $f(t) = 12$ gives the value for k. $12 = \dfrac{200}{1 + 32e^{-3k}}$, $12 \cdot 32e^{-3k} = 188$,

$e^{-3k} = 0.49$, $\ln 0.49 = -3k$, $k = 0.24$. The specific formula is $f(t) =$

$\dfrac{200}{1 + 32e^{-0.24t}}$, and the number of parlors in 1996, $t = 6$ is $f(6) =$

$\dfrac{200}{1 + 32e^{-1.44}} = 23$

12. Take dq/dt and evaluate at 10 months. $\dfrac{dq}{dt} = \dfrac{0 - 5000(2)(-0.4)e^{-0.4t}}{(1 + 2e^{-0.4t})^2}$ and

$\left.\dfrac{dq}{dt}\right|_{10} = 70$ games per month. Depending on where you round off you may get a number slightly different from this.

CHAPTER 10

◆◆◆◆◆◆◆◆◆◆◆◆◆◆◆◆◆◆◆◆◆◆◆◆◆◆◆◆◆◆◆◆◆◆◆◆

MORE INTEGRALS

◆◆◆◆◆◆◆◆◆◆◆◆◆◆◆◆◆◆◆◆◆◆◆◆◆◆◆◆◆◆◆◆◆◆◆◆

Do I Need to Read This Chapter?   You should read this chapter if you need to review or you need to learn about

➡ Volumes

➡ Arc lengths

➡ Surfaces of revolution

➡ Change of variable

➡ Integration by parts

➡ Partial fractions

➡ Approximation methods

There is an almost limitless supply of increasingly complex integrals and applications of integrals. Depending on your interests, certain topics and integrals in this chapter may be very interesting to you while others may be completely uninteresting.

If you are looking for help with a particular integral or a particular application, you may want to survey the chapter to find those problems and associated discussions that fit your interest. We have concentrated on four topics: volumes, arc lengths, surface areas, and non-standard integrals that occur often in real-world problems. This latter area is often called techniques of integration, the name suggesting the study of integration techniques that work for a number of different problems. Most texts and extensive integral tables are organized around various categories of integrals. Within the space limitations, we have attempted to pick those integrals and applications that will help the largest number of people. We start with a discussion of volumes.

10-1   Volumes

Finding volumes of non-standard geometric shapes can only be accomplished with calculus. This work is a logical extension of the study of the calculation of areas using calculus. You will find many parallels between area and volume calculations. Finding volumes is also extremely visual. If you can visualize the problem, you can usually do it. We start with some simple problems and work up to the more challenging ones. The first problem uses the method of disks to calculate the volume generated by rotating a parabola of a fixed height about its symmetry axis. Next the problem is done again using the method of cylindrical shells.

Example 10-1   Find the volume generated by rotating $y = x^2$ about the y-axis and bounded by the plane $y = 4$.

In two dimensions, $y = 4$ defines a horizontal line at $y = 4$. In three dimensions, $y = 4$ defines a plane normal to the y-axis and parallel to the x-z plane.

Solution: This is a three-dimensional picture. Start with the $y = x^2$ curve from $y = 0$ up to $y = 4$ (and $x = \pm 2$). The rotation of this part of the parabola about the y-axis produces a rounded cone shape (Fig. 10-1).

The volume can be viewed as a collection (integral?) of disks of width dy and radius dictated by the radius of the cone. The volume of each of these disks is generically $\pi \times (radius)^2 \times thickness$. The radius of the disk is x so the differential volume of each disk can be written $\pi x^2 dy$. The sum of all these disks is an integral over y.

Start by writing $\int \pi x^2 dy$.

The first thing wrong with this integral is the x^2 term. If the integral is over dy, we can't have xs under the integral sign. Replace x^2 by its equivalent, y.

The next thing that needs to be added to the integral is the limits. There are none. Integration in the y-direction is from $y = 0$ to $y = 4$. The curve starts at $y = 0$, and the problem gives the upper boundary as the plane $y = 4$.

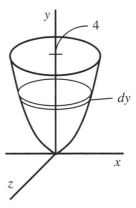

Fig. 10-1

The volume integral is $V = \pi \int_0^4 y \, dy = \pi \left. \dfrac{y^2}{2} \right|_0^4 = 8\pi$.

Example 10-2 Find the volume generated by rotating $y = x^2$ about the y-axis and bounded by the plane $y = 4$ using the method of cylindrical shells.

Solution: In the previous problem the volume was visualized as a stack of disks of thickness dy. This is the method of disks.

This problem prescribes using the method of cylindrical shells. Visualize a cylinder, actually a cylindrical shell, of radius x, height the difference between $y = 4$ and $y = x^2$ and width dx.

The volume of the cylindrical shell, as shown in Fig. 10-2, is 2π times the radius times the height of the shell times the thickness of the shell. The 2π times the radius effectively wraps the rectangle of height between the curves and width dx around the y-axis.

The radius is x, the height of the rectangle is $(4 - x^2)$ (the top of the rectangle is at $y = 4$ and the bottom of the rectangle is on the $y = x^2$ curve), and the width is dx, so the differential volume is $2\pi x(4 - x^2)dx$. The sum of these cylindrical shells is an integral over x.

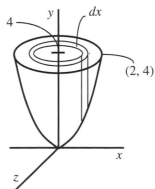

Start by writing an integral

$$\int 2\pi x(4 - x^2)dx$$

Fig. 10-2

The integral is in x from 0 to 2 so

$$V = \int_0^2 2\pi x(4 - x^2)dx = 2\pi \int_0^2 (4x - x^3)dx = 2\pi\left[\frac{4x^2}{2} - \frac{x^4}{4}\right]_0^2$$

$$= 2\pi[8 - 4] = 8\pi$$

Example 10-3 Find the volume generated by rotating the area bounded by $x^2 + y^2 = 25$, $x = 5$, and $y = 4$ about the x-axis.

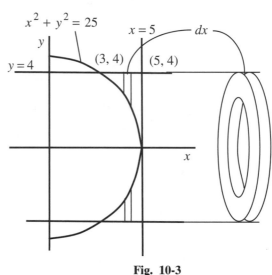

Solution: Start by finding the area to be rotated. The line $x = 5$ and the circle intersect at $x = 5$ on the x-axis. The line $y = 4$ intersects the circle when $y = 4$ ($x^2 + 4^2 = 25$, $x^2 = 9$, $x = 3$). The circle and the line $y = 4$ intersect at (3, 4).

Fig. 10-3

Visualize the volume obtained by rotating this area about the x-axis as composed of disks with outer radius equal to 4, inner radius on the circle, and width dx. See Fig. 10-3. The outer radius of the disk is 4 and the inner radius is the solution of $x^2 + y^2 = 25$ for y ($x^2 + y^2 = 25$, $y = \sqrt{25 - x^2}$). The differential volume of the disk is $\pi[$(outer radius)^2 minus (inner radius)$^2]$ times dx. The integral in dx is from 3 to 5, so the volume integral is

$$V = \int_3^5 \pi[4^2 - (25 - x^2)]dx = \pi \int_3^5 (16 - 25 + x^2)dx = \pi \int_3^5 (-9 + x^2)dx$$

$$= \pi\left[-9x + \frac{x^3}{3}\right]_3^5$$

$$V = \pi\left\{\left[-45 + \frac{125}{3}\right] - [-27 + 9]\right\} = \pi\left\{-45 + \frac{125}{3} + 27 - 9\right\}$$

$$= \pi\left\{\frac{125}{3} - 27\right\}$$

$$V = \pi\left\{\frac{125 - 81}{3}\right\} = \frac{44\pi}{3}$$

Example 10-4 Find the volume of cement required to build the top of a bird-bath. The bottom of the birdbath follows the parabola $y = 0.1x^2$. The inside of the birdbath follows $y = 0.20 + 0.08x^2$. All the dimensions are in feet. The top edge of the birdbath is bounded by the horizontal line $y = 0.4$.

Solution: Start with the profile of the birdbath in x-y as shown in Fig. 10-4. The bottom parabola, $y = 0.1x^2$, starts at $x = 0$, $y = 0$, and intersects $y = 0.4$ when $x = 2$ $(0.4 = 0.1x^2, x^2 = 4, x = 2)$.

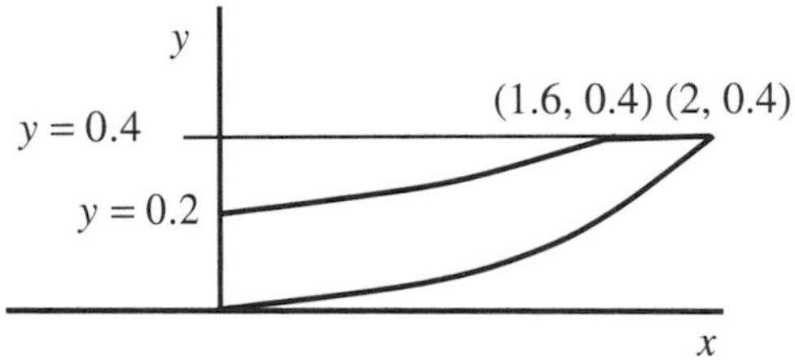

The top parabola starts at $x = 0$, $y = 0.2$ and intersects $y = 0.4$ when $x = 1.6$

$(0.4 = 0.2 + 0.08x^2, 0.2 = 0.08x^2, x = 1.6)$.

Fig. 10-4

Now rotate the profile around the y-axis. The volume of the birdbath is the volume inside the bottom parabola up to $y = 0.4$ minus the volume inside the top

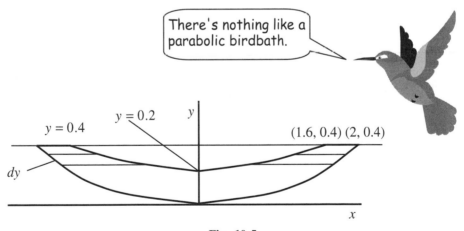

Fig. 10-5

parabola up to $y = 0.4$. Figure 10-5 shows the birdbath and the disks. The volume of the disks is π times (radius)^2 times thickness.

The volume within the bottom parabola is $\displaystyle\int_0^{0.4} \pi (x_{bottom})^2 dy = \pi \int_0^{0.4} \frac{y}{0.1}\, dy.$

The volume within the inner parabola is $\int_{0.2}^{0.4} \pi(x_{inner})^2 dy = \pi \int_{0.2}^{0.4} \dfrac{y - 0.2}{0.08} dy$

Be careful of the limits!

The volume of cement in the birdbath is the larger volume minus the smaller volume as shown by the disks in Fig. 10-5. The volume integral is

$$V = \dfrac{\pi}{0.1} \int_0^{0.4} y\, dy - \dfrac{\pi}{0.08} \int_{0.2}^{0.4} (y - 0.2) dy = \dfrac{\pi}{0.1} \left\{ \dfrac{y^2}{2} \right\}_0^{0.4} - \dfrac{\pi}{0.08} \left\{ \dfrac{y^2}{2} - 0.2y \right\}_{0.2}^{0.4}$$

$$V = \dfrac{\pi}{0.1} \left\{ \dfrac{0.16}{2} \right\} - \dfrac{\pi}{0.08} \left\{ \left[\dfrac{0.16}{2} - 0.08 \right] - \left[\dfrac{0.04}{2} - 0.04 \right] \right\}$$

$$V = \dfrac{\pi}{0.1} \{0.08\} - \dfrac{\pi}{0.08} \{[0.08 - 0.08] - [0.02 - 0.04]\}$$

$$= 0.8\pi - \dfrac{\pi}{0.08} \{0 + 0.02\}$$

$$V = 0.8\pi - \dfrac{\pi}{4} = \pi(0.8 - 0.25) = 0.55\pi = 1.73 \text{ ft}^3 \text{ of cement}$$

10-2   Arc Lengths

A small length of a curve in x-y denoted by ds can be written in terms of dx and dy using Pythagorean theorem. The geometry of ds, dx, and dy are shown in Fig. 10-6 and the Pythagorean relation is

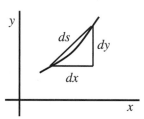

Fig. 10-6

$$ds^2 = dx^2 + dy^2$$

Any small change in s can be viewed as a small change in x and a small change in y. Solve this equation for ds

$$ds = \sqrt{dx^2 + dy^2}$$

and factor out first a dx, and then a dy.

$$ds = \sqrt{1 + \left(\dfrac{dy}{dx} \right)^2} dx = \sqrt{1 + \left(\dfrac{dx}{dy} \right)^2} dy$$

This little exercise is sufficiently easy so that you do not have to use precious memory space remembering it, just work it out as needed. The total length of an arc is the integral between the appropriate limits of this differential statement.

$$\int ds = \int \sqrt{1 + \left(\frac{dy}{dx}\right)^2}\, dx = \int \sqrt{1 + \left(\frac{dx}{dy}\right)^2}\, dy$$

Because of the square root, and the square of the slope, the integrals are usually not easy.

The curve $y^2 = x^3$ turns out to be one of the easier arc lengths to calculate. Form dy/dx.

$$2y\,dy = 3x^2 dx \quad \text{and} \quad \frac{dy}{dx} = \frac{3x^2}{2y} \quad \text{so} \quad \left(\frac{dy}{dx}\right)^2 = \frac{9x^4}{4y^2} = \frac{9x^4}{4x^3} = \frac{9x}{4}$$

The general integral for arc length of this curve is $s = \int \sqrt{1 + \frac{9x}{4}}\, dx$.

Most of the time the integrals are so difficult it is worth looking at both formulas for the arc length in an attempt to find the easiest integral. The other possible integral starts from dx/dy.

$$\frac{dx}{dy} = \frac{2y}{3x^2} \quad \text{and} \quad \left(\frac{dx}{dy}\right)^2 = \frac{4y^2}{9x^4}$$

The x^4 term in the denominator cannot be conveniently written in terms of y without getting into fractional powers so the previous integral looks at this point to be the easier. This type of integral will be taken up in an example later.

Example 10-5 Find the length of arc between $x = 0$ and $x = 1$ for the curve $y = x^{3/2}$.

Solution: First find dy/dx and form the two square roots to see which looks easier.

$$dy = \frac{3}{2}x^{1/2}dx \quad \Rightarrow \quad \frac{dy}{dx} = \frac{3}{2}x^{1/2} \quad \Rightarrow \quad ds = \sqrt{1 + \frac{9x}{4}}\, dx$$

$$\Rightarrow \quad ds = \sqrt{1 + \frac{4}{9x}}\, dy$$

The integral in dx is easier so set up the integral with limits.

$$S = \int_0^1 \sqrt{1 + \frac{9x}{4}}\, dx$$

A change of variable should help: let $u = 1 + 9x/4$ and $du = (9/4)dx$.

$$S = \int_1^{13/4} u^{1/2} \frac{4}{9}\, du = \frac{4}{9}\frac{2}{3} u^{3/2}\Big|_1^{13/4} = \frac{8}{27}\Big[\Big(\frac{13}{4}\Big)^{3/2} - 1^{3/2}\Big] = 1.44$$

In this case, with a complicated change of variable it looked easier to change the limits on the integral than convert back to x's.

10-3   Surfaces of Revolution

Determining the surface area of non-standard shapes is another uniquely calculus problem. The technique for finding the surface area of a shape produced by rotating a curve is similar to finding volumes and additionally uses concepts from length of arc calculations.

Start with a parabola, $y = x^2$, rotated about the y-axis and consider the surface of that parabola up to $y = 4$. The curve doesn't have to be a parabola. A parabola is just convenient to visualize. The surface area is viewed as a collection of strips wrapped around the parabola. The area of these strips is 2π (radius), the length around, times the width of the strip, ds. The differential piece of surface for a curve rotated about the y-axis is $2\pi x ds$ (Fig. 10-7).

The ds is $\sqrt{1 + \Big(\dfrac{dy}{dx}\Big)^2}\, dx$ or $\sqrt{1 + \Big(\dfrac{dx}{dy}\Big)^2}\, dy$.

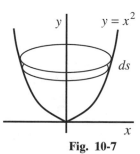

Fig. 10-7

Since the length of the strip ds is written $2\pi x$ look first at the

$$\sqrt{1 + \Big(\frac{dy}{dx}\Big)^2}\, dx \text{ form for } ds$$

Start with $\dfrac{dy}{dx} = 2x$ and $\Big(\dfrac{dy}{dx}\Big)^2 = 4x^2$ and

$\sqrt{1 + \Big(\dfrac{dy}{dx}\Big)^2} = \sqrt{1 + 4x^2}$. This looks to be the best form for ds, because it does not contain a fraction inside the square root.

The integral for the surface area of the parabola then is

$$A = 2\pi \int_0^2 x\sqrt{1 + 4x^2}\,dx$$

The integrals encountered in surface area calculations are usually worse than the ones for arc length. The next section will start on techniques of integration and this integral will be evaluated there.

Example 10-6   Find the area generated by rotating the curve $y = (3x)^{1/3}$ about the y-axis from $y = 0$ to $y = 2$.

Solution: Sketch the curve (Fig. 10-8). Take the derivative and look for which of the two integrals that looks easiest.

Rewrite the equation of the curve as $y^3 = 3x$.

Then $dx = \frac{1}{3}3y^2dy = y^2dy$ and $\frac{dx}{dy} = y^2$. The area inte-

gral in dy looks to be the easiest. $ds = \sqrt{1 + \left(\frac{dx}{dy}\right)^2}\,dy$ so

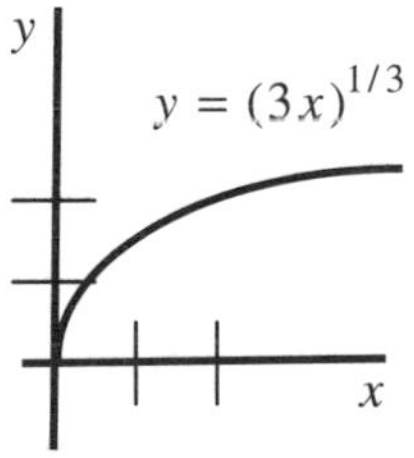

Fig. 10-8

$$A = 2\pi \int_0^2 x\sqrt{1 + y^4}\,dy = \frac{2\pi}{3} \int_0^2 y^3\sqrt{1 + y^4}\,dy$$

Now make a change of variable with $u = 1 + y^4$ and $du = 4y^3dy$.

$$A = \frac{2\pi}{3}\frac{1}{4} \int_1^{17} \sqrt{u}\,du = \frac{\pi}{6}\frac{2}{3} u^{3/2}\Big|_1^{17} = \frac{\pi}{9}[(17)^{3/2} - 1] = 24$$

10-4   Techniques of Integration

In this section we show you some techniques for handling particularly difficult looking integrals. Along the way some interesting practical problems that so far have been avoided because of the difficulty of integrating will be done. These techniques of integration are categorized by technique. We start with the simpler techniques and work through the more popular, or more often encountered.

Change of Variable

The change of variable technique is also called the method of substitution. As the names imply the approach is to define a new variable that will transform the integral to one that is a standard form. The procedure is to define a new variable, take the derivative of that new variable and then write the integral in terms of the new variable and derivative. There is some skill in picking the new variable and sometimes you just have to try a few. The best first guess for a change of variable is to look for the worst looking part of the integral and make that worst looking part the new variable or at least incorporate it into the new variable. The best way to understand any of these techniques is to jump right in and start doing some problems.

Example 10-7   Find $\int xe^{x^2}dx.$

Solution: Make a change of variable. Let $u = x^2$ so that $du = 2xdx$. This transforms the integral.

$$\int xe^{x^2}dx = \frac{1}{2}\int e^u du = \frac{1}{2}e^u = \frac{1}{2}e^{x^2}$$

Remember to translate back to the original variable at the end of the problem.

Example 10-8   Find $\int \frac{dx}{x\ln x}.$

Solution: The worst looking part of this integral is the $\ln x$ so make a substitution $v = \ln x$ so that $dv = \frac{1}{x}dx$. This transforms the integral into a standard form.

$$\int \frac{dx}{x\ln x} = \int \frac{1}{v}dv = \ln v = \ln(\ln x)$$

This is a bit of a strange answer, but then it was a bit of a strange integral.

Example 10-9   Find $\int \sqrt{3 - 2x}dx.$

Solution: A new variable $3 - 2x$ would allow the $\sqrt{3 - 2x}$ to be written as a power, and integrals of a variable raised to a power d (variable) are standard integrals. Let $w = 3 - 2x$ with $dw = -2dx$. The integral is transformed and solved.

$$\int \sqrt{3 - 2x}dx = -\frac{1}{2}\int w^{1/2}dw = -\frac{1}{2}\frac{w^{3/2}}{3/2} = -\frac{1}{3}w^{3/2} = -\frac{1}{3}(3 - 2x)^{3/2}$$

Example 10-10 Find the arc length between $x = 1$ and $x = 3$ for the curve $y^2 = x^3$.

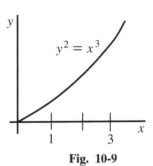

Fig. 10-9

Solution: This is the curve used as the example in the discussion of arc lengths (previous section). A rough sketch of the curve is shown in Fig. 10-9. Taking the square root of both sides the $y^2 = x^3$ equation produces $y = x^{1.5}$.

This is a curve that has a shape somewhere between the shape of $y = x$, a straight line, and $y = x^2$, a parabola. The curve slopes upward but not as rapidly as the quadratic.

The general formula for the arc length is $S = \int_1^3 \sqrt{1 + \left(\dfrac{dy}{dx}\right)^2}\, dx$.

For the curve $y^2 = x^3$, $\dfrac{dy}{dx} = \dfrac{3x^2}{2y}$ and $\left(\dfrac{dy}{dx}\right)^2 = \dfrac{9x^4}{4y^2} = \dfrac{9x^4}{4x^3} = \dfrac{9x}{4}$ so the arc length from $x = 1$ to $x = 3$ is

$$S = \int_1^3 \sqrt{1 + \frac{9x}{4}}\, dx = \int_1^3 \sqrt{\frac{4 + 9x}{4}}\, dx = \frac{1}{2}\int_1^3 \sqrt{4 + 9x}\, dx$$

The integral has been simplified to the point where a change of variable is in order. Let $u = 4 + 9x$ so that $du = 9dx$ and rewrite the integral.

$$S = \frac{1}{2}\int_1^3 \sqrt{4 + 9x}\,dx = \frac{1}{2}\frac{1}{9}\int_{x=1}^{x=3} u^{1/2}du = \frac{1}{18}\frac{u^{3/2}}{3/2}\bigg|_{x=1}^{x=3} = \frac{1}{27}(4 + 9x)^{3/2}\bigg|_{x=1}^{x=3}$$

$$S = \frac{1}{27}\{[4 + 27]^{3/2} - [4 + 9]^{3/2}\} = \frac{1}{27}[31^{3/2} - 13^{3/2}]$$

$$= \frac{1}{27}[172.6 - 46.9] = 4.66$$

The limits on the integral can be confusing. The strictly correct way to evaluate the integral is to change the limits when the variable is changed. Looking at the definition of u; for $x = 1$, $u = 4 + 9(1) = 13$, and for $x = 3$, $u = 4 + 9(3) = 31$. Using this approach the integrals would read

$$S = \frac{1}{2}\int_1^3 \sqrt{4 + 9x}\,dx = \frac{1}{2}\frac{1}{9}\int_{13}^{31} u^{1/2}du = \frac{1}{18}\frac{u^{3/2}}{3/2}\Big|_{13}^{31} = \frac{1}{27}[31^{3/2} - 13^{3/2}] = 4.66$$

However you choose to do the problem be careful of the limits. If you write $x = \ldots$ or $u = \ldots$ in the limits you will avoid getting confused. There are enough pitfalls in evaluating these integrals without getting tripped up with the limits.

Example 10-11 Find the surface area of $y = x^2$ rotated about the y-axis from $x = 0$ to $x = 2$.

Solution: The surface area is the area generated by rotating the parabola about the y-axis up to $y = 4$, which corresponds to $x = 2$. This is the problem used to illustrate the calculation of surface areas. The integral is the integral of a strip of surface area with length equal to the circumference, 2π(radius) times the differential length along any arc of the surface, ds.

Figure 10-10 shows the parabola up to $y = 4$ corresponding to $x = 2$ and the differential strip of area. The differential area is

$$dA = 2\pi x\,ds = 2\pi x\sqrt{1 + \left(\frac{dy}{dx}\right)^2}\,dx$$

and $\dfrac{dy}{dx}$ of $y = x^2$ is $2x$ so

$$A = 2\pi\int_0^2 x\sqrt{1 + 4x^2}\,dx$$

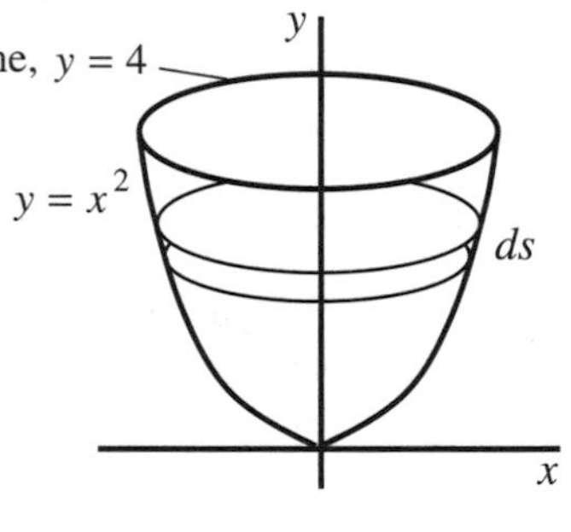

Fig. 10-10

In looking for a change of variable, look for the worst part of the integral, which is the $1 + 4x^2$. Let $v = 1 + 4x^2$ with $dv = 8x\,dx$. Replace $1 + 4x^2$ with v and $x\,dx$ with $dv/8$. Change the limits. When $x = 0$, $v = 1$, and when $x = 2$, $v = 17$.

$$A = 2\pi\int_0^2 x\sqrt{1 + 4x^2}\,dx = 2\pi\frac{1}{8}\int_1^{17} v^{1/2}dv$$

$$A = \frac{\pi}{4}\left[\frac{v^{3/2}}{3/2}\right]_1^{17} = \frac{\pi}{6}[17^{3/2} - 1^{3/2}] = \frac{\pi}{6}[70.1 - 1] = 36.2$$

Example 10-12   Find $\int \tan(3x - 2)dx$.

Solution: The worst part of this integral is the $3x - 2$ so let $w = 3x - 2$ and $dw = 3dx$. The integral transforms to a standard integral.

$$\int \tan(3x - 2)dx = \frac{1}{3}\int \tan w \, dw = -\frac{1}{3}\ln(\cos w) = -\frac{1}{3}\ln[\cos(3x - 2)]$$

Example 10-13   The price of a product varies with supply and demand in such a way that $dp/dt = k(5 - 2p)$ with k to be determined by conditions. Find the price as a function of time and graph the price vs. time. The price is \$4.50 when $t = 0$, and \$4.00 when $t = 2$. The t is in years.

Solution: The first step in solving for $p(t)$ is to write the rate statement in a form that can be integrated.

$$\frac{dp}{5 - 2p} = kdt \quad \text{or} \quad \int \frac{dp}{5 - 2p} = \int kdt$$

Deal with this integral in p as a separate exercise. Make a substitution for $5 - 2p$ by letting $z = 5 - 2p$ so $dz = -2dp$ and the integral becomes

$$\int \frac{dp}{5 - 2p} = -\frac{1}{2}\int \frac{dz}{z} = -\frac{1}{2}\ln z = -\frac{1}{2}\ln(5 - 2p)$$

With this little side calculation and remembering that integrating $\int \frac{dp}{5 - 2p} = \int kdt$ produces a constant of integration, the integration produces

$$-\frac{1}{2}\ln(5 - 2p) = kt + C$$

Rearranging for convenience in writing as an exponent: $\ln(5 - 2p) = -2kt - 2C$. And writing as an exponential (This is the only way to get an equation that reads $P = \cdots$) we get

$$5 - 2p = e^{-2kt-2C} = e^{-2kt}e^{-2C}$$

The constant of integration can be carried as long as you like but defining a new constant at this point looks convenient. Make e^{-2C} equal to $2D$.

$$-2p = -5 + 2De^{-2kt}$$
$$p = 2.5 - De^{-2kt}$$

Now apply the condition that $p = 4.5$ at $t = 0$.

$$4.5 = 2.5 - D(1) \quad \text{and} \quad -D = 2$$

so

$$p = 2.5 + 2e^{-2kt}$$

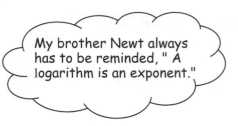

The second condition that $p = 4$ when $t = 2$ will define the constant k.

$$4 = 2.5 + 2e^{-4k}, \ 1.5 = 2e^{-4k},$$

$$0.75 = e^{-4k}$$

Switch to logarithms:

$$-4k = \ln 0.75,$$

$$k = -\frac{1}{4} \ln 0.75 = 0.072,$$

and $\qquad 2k = 0.144$

Finally,

$$p = 2.5 + 2e^{-0.144t}$$

Now graph the function. At $t = 0$, $p = 4.5$ as given in the problem. As time goes on, the $2e^{-0.144t}$ term gets smaller and smaller and as $t \to \infty$, $p \to 2.5$. The line $p = 2.5$ is an asymptote. The curve is shown in Fig. 10-11.

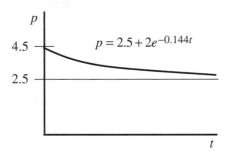

Fig. 10-11

Trigonometric Integrals

There are a large, large number of trigonometric integrals. Some are relatively easy. Most are relatively difficult. Solving trigonometric integrals involves changes of variables and using trigonometric identities and a good bit of ingenuity, imagination some might call it. The following several problems demonstrate the more popular techniques (did someone say tricks?) for solving trigonometric integrals.

Example 10-14 Find $\int \sin^2 x \cos x \, dx$.

Solution: Recognizing that $\cos x$ is the derivative of $\sin x$ suggests a change of variable might make this integral into a standard form. Take $u = \sin x$ and $du = \cos x \, dx$. Making these substitutions

$$\int \sin^2 x \, \cos x \, dx = \int u^2 du = \frac{u^3}{3} = \frac{\sin^3 x}{3}$$

Example 10-15 Find $\int (\sin^3 x)(\cos^3 x) dx$.

Solution: This problem is a little harder than the previous one. If we let $v = \sin x$ and $dv = \cos x \, dx$, the integral becomes

$$\int (\sin^3 x)(\cos^3 x) dx = \int v^3(\cos^2 x) dv$$

which doesn't seem to be much of an improvement. However, using the identity $\sin^2 x + \cos^2 x = 1$, $\cos^2 x = 1 - \sin^2 x$ and the integral becomes

$$\int v^3(1 - v^2) dv = \int (v^3 - v^5) dv = \frac{v^4}{4} - \frac{v^6}{6} = \frac{v^4}{4}\left(1 - \frac{2v^2}{3}\right)$$

$$= \frac{\sin^4 x}{4}\left(1 - \frac{2\sin^2 x}{3}\right)$$

Example 10-16 Find $\int \sqrt{\sin x}(\cos^3 x) dx$.

Solution: Use the identity $\sin^2 x + \cos^2 x = 1$ to replace $\cos^2 x$ so the integral now reads

$$\int \sqrt{\sin x}\,(1 - \sin^2 x) \cos x \, dx$$

Now make a change of variable. Let $w = \sin x$ and $dw = \cos x \, dx$ so that the integral now reads

$$\int w^{1/2}(1 - w^2) dw = \int (w^{1/2} - w^{5/2}) dw = \frac{w^{3/2}}{3/2} - \frac{w^{7/2}}{7/2}$$

$$= 2\left[\frac{(\sin x)^{3/2}}{3} - \frac{(\sin x)^{7/2}}{7}\right]$$

All trigonometric integrals are not this easy. Though there are some patterns to doing trigonometric integrals, as demonstrated in the previous problems, trigonometric integrals can be some of the most difficult you will encounter. Fortunately, there are tables of trigonometric integrals that will help you out of most problems.

10-5   Integration by Parts

Integration by parts is somewhat similar to the method of substitution in that the correct association will make a difficult integral into a not so difficult integral. The formula for integration by parts, which we will not derive or even justify, is found in the Mathematical Tables.

$$\int u\,dv = uv - \int v\,du$$

The key to successful application of this rule is the correct initial choice of u and dv. Sometimes you have to try more than one combination to get one to work well. The purpose of the choice is to make the integral on the right side easier and not harder than the one you started with. The best way to learn this is to go directly to some problems and see how it is done.

Example 10-17   Find $\int xe^x dx$.

Solution: Fit the integral to the pattern $\int u\,dv = uv - \int v\,du$.

A good first identification is to take $e^x dx$ as dv, and x as u. If this identification is made then $du = dx$ and $\int e^x dx = \int dv$ makes $e^x = v$. Follow the pattern and write

$$\int u(dv) = uv - \int v\,du$$

$$\int x(e^x dx) = xe^x - \int e^x dx$$

The integral $\int e^x dx$ is e^x so

$$\int xe^x dx = xe^x - e^x = e^x(x - 1)$$

Example 10-18 Find $\int x^2 e^x dx$ by integration by parts.

Solution: The form of integration by parts is $\int u(dv) = uv - \int v du$. Take $u = x^2$ and $dv = e^x dx$. From these identifications $du = 2x dx$ and $\int dv = \int e^x dx$ makes $v = e^x$. Write the original integral as an integration by parts.

$$\int x^2 e^x dx = x^2 e^x - 2 \int x e^x dx$$

The $\int x e^x dx$ can itself be integrated by parts as was done in the previous problem. Use the result of Example 10-17 to write

$$\int x^2 e^x dx = x^2 e^x - 2(x e^x - e^x) = x^2 e^x - 2x e^x + 2e^x = e^x(x^2 - 2x + 2)$$

This problem is an excellent example of the multiple uses of integration by parts. Multiple integrations by parts are typical of complex exponential and trigonometric integrals.

Example 10-19 Find $\int \ln x dx$ by integration by parts.

Solution: Let $u = \ln x$ and $dv = dx$, so $du = \dfrac{dx}{x}$ and $v = \int dx = x$ so that

$$\int u dv = uv - \int v du$$

$$\int \ln x dx = x \ln x - \int x \frac{dx}{x} = x \ln x - x + C$$

Example 10-20 The income for a certain company is a combination of steady growth and a cyclic component with the income following $S = 2t + t \sin(\pi t/2)$, where S is in tens of thousands of dollars per month and t is a quarter of a year ($t = 1$ corresponds to 3 months, or one-quarter). The income for any period is the integral of this income per month function over that period. Find the income for the next 3 quarters.

Solution: The income for 3 quarters is the integral of S over t from $t = 0$ to $t = 3$.

$$S = \int_0^3 [2t + 3t \sin(\pi t/2)] \, dt = 2 \int_0^3 t \, dt + \left(\frac{2}{\pi}\right)^2 \int_{t=0}^{t=3} \left(\frac{\pi t}{2}\right) \sin\left(\frac{\pi t}{2}\right) d\left(\frac{\pi t}{2}\right)$$

Take the first integral as S_1 and the second integral as S_2. The first integral $S_1 = \dfrac{2t^2}{2}\Big|_{t=0}^{t=3} = 9$ and means that $90,000 in income was received in the 3 quarters (S is in tens of thousands of dollars).

Make a change of variable in the second equation. Let $y = \pi t/2$ so $dy = d(\pi t/2)$. The new limits are: for $t = 0$, $y = 0$ and for $t = 3$, $y = 3\pi/2$. The second integral now reads

$$S_2 = \left(\frac{2}{\pi}\right)^2 \int_{y=0}^{y=3\pi/2} y \sin y \, dy$$

The integration is performed by parts.

$$\int u \, dv = uv - \int v \, du$$

Let $u = y$ and $dv = \sin y \, dy$ producing $du = dy$ and $\int dv = \int \sin y \, dy$ or $v = -\cos y$.

$$S_2 = \left(\frac{2}{\pi}\right)^2\left[-y\cos y - \int(-\cos y)\, dy\right]_{y=0}^{y=3\pi/2} = \left(\frac{2}{\pi}\right)^2\left[-y\cos y + \sin y\right]_{y=0}^{y=3\pi/2}$$

$$S_2 = \left(\frac{2}{\pi}\right)^2\left\{\left[-\frac{3\pi}{2}\cos\left(\frac{3\pi}{2}\right) + \sin\frac{3\pi}{2}\right] - \left[-(0)\cos 0 + \sin 0\right]\right\}$$

The $\sin 0 = 0$ and $(0)\cos 0 = 0$ so the second bracket is zero.

The $\cos\dfrac{3\pi}{2} = 0$ and $\sin\dfrac{3\pi}{2} = -1$ so this integral is

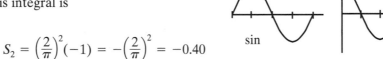

sin   cos

$$S_2 = \left(\frac{2}{\pi}\right)^2(-1) = -\left(\frac{2}{\pi}\right)^2 = -0.40$$

This means a loss due to this cyclic component of $4000. The total income over the 3 quarters is $90,000 minus $4000 or $86,000.

10-6   Partial Fractions

A single complicated fraction can often be written as two fractions, each of which is less complicated than the original.

The fraction

$$\frac{x^2 + 2x + 1}{x^3} \text{ can be written as } \frac{x^2}{x^3} + \frac{2x}{x^3} + \frac{1}{x^3} \text{ or as } \frac{1}{x} + \frac{2}{x^2} + \frac{1}{x^3}$$

If you needed to integrate this fraction it would be much easier to integrate three simpler fractions, than the more complicated single fraction. Making multiple simpler fractions from a single fraction is a logical process that is best learned by working an example.

To use partial fractions the denominator has to be factorable and it is most convenient if the numerator is one degree less than the denominator. If the numerator and denominator are the same degree or the numerator is a higher degree than the denominator, then long divide to reduce the fraction. The first partial fractions to consider all have nonrepeating and nonquadratic factors in their denominators.

Example 10-21   Write $\dfrac{2}{x^2 - x - 2}$ in partial fractions.

Solution: The denominator of the fraction can be factored $x^2 - x - 2 = (x - 2) \times (x + 1)$ indicating that the fraction can be written as something over the first factor and something else over the second factor. Set up two fractions with undetermined numerators and equal to the original fraction.

$$\frac{2}{(x - 2)(x + 1)} = \frac{A}{x - 2} + \frac{B}{x + 1}$$

As with most equations involving fractions multiply both sides by the common denominator to clear the fractions.

$$2 = A(x + 1) + B(x - 2) = Ax + A + Bx - 2B$$
$$= (A + B)x + (A - 2B)$$

Equating the constants and the coefficients of x produces two identities: $A + B = 0$ and $A - 2B = 2$. This is sufficient information to determine A and B. Subtract the second identity from the first $[(A + B = 0) - (A - 2B = 2)]$ to eliminate A. Now $3B = -2$ or $B = -2/3$. If $B = -2/3$, then $A = 2/3$ $(A + B = 0)$ and the original fraction is now written as

$$\frac{2}{(x - 2)(x + 1)} = \frac{2/3}{(x - 2)} - \frac{2/3}{(x + 1)}$$

Example 10-22 Integrate $\displaystyle\int \frac{3x^2 + 7x - 4}{x^3 - 4x}$ by partial fractions.

Solution: First write the fraction in terms of partial fractions. The denominator is factorable so write

$$\frac{3x^2 + 7x - 4}{x(x + 2)(x - 2)} = \frac{A}{x} + \frac{B}{x + 2} + \frac{C}{x - 2}$$

Multiply by the common denominator.

$$3x^2 + 7x - 4 = A(x^2 - 4) + B(x^2 - 2x) + C(x^2 + 2x)$$

$$= (A + B + C)x^2 + (-2B + 2C)x + (-4A)$$

Write the identities
$$
\begin{array}{l}
A + B + C = 3 \\
-2B + 2C = 7 \\
-4A = -4
\end{array}
$$
. With $A = 1$,
$$
\begin{array}{l}
B + C = 2 \\
-2B + 2C = 7
\end{array}
$$

Multiply the first equation by 2 and add the equations to eliminate B so $4C = 11$ or $C = 11/4$. Substitute in $B + C = 2$ so $B = 2 - 11/4 = 8/4 - 11/4 = -3/4$. The fraction now is written as

$$\frac{3x^2 + 7x - 4}{x^3 - 4x} = \frac{1}{x} - \frac{3/4}{x + 2} + \frac{11/4}{x - 2}$$

The integral now reads

$$\int \frac{3x^2 + 7x - 4}{x^3 - 4x} = \int \frac{1}{x}\, dx - \frac{3}{4}\int \frac{1}{x + 2}\, dx + \frac{11}{4}\int \frac{1}{x - 2}\, dx$$

The integral $\displaystyle\int \frac{1}{x - a}\, dx$ is a logarithmic derivative.

Replace $x - a$ with u and $du = dx$ so $\displaystyle\int \frac{1}{x - a}\, dx$ is in the form $\displaystyle\int \frac{1}{x}\, dx$.

The three integrals can now be written easily.

$$\int \frac{3x^2 + 7x - 4}{x^3 - 4x} = \ln x - \frac{3}{4}\ln(x + 2) + \frac{11}{4}\ln(x - 2)$$

The next complication in partial fractions is when there is a repeating factor in the denominator. This requires two partial fractions, one with the factor to the first power and a second with the factor to the second power. If the denominator contains multiple identical factors, then make as many fractions as there are factors. And the final complication is when there is a quadratic factor in the denominator. A quadratic factor requires a numerator in the fraction that is a constant plus another constant times the variable. This is best illustrated with the next two examples.

Example 10-23   Find the $\int \dfrac{x^2 - 4x - 1}{(x - 1)^2}\, dx$ using partial fractions.

Solution: The partial fraction is written as

$$\frac{x^2 - 4x - 1}{(x - 1)^2} = \frac{A}{(x - 1)} + \frac{B}{(x - 1)^2}$$

Multiply by the common denominator and compare coefficients.

$$x^2 - 4x - 1 = A(x - 1) + B = Ax - A + B$$

$A = -4, B - A = -1, B = A - 1 = -5$, so the integral is written as

$$\int \frac{x^2 - 4x - 1}{(x - 1)^2}\, dx = -\int \frac{4}{x - 1}\, dx - \int \frac{5}{(x - 1)^2}\, dx$$

$$= -4 \ln (x - 1) + 5(x - 1)^{-1} + C$$

Example 10-24   Find $\int \dfrac{5x^3 - 3x^2 + 2x - 1}{x^4 + x^2}\, dx$ using partial fractions.

Solution: When the denominator is factored there is a repeating term plus a quadratic term, that is, one that cannot be factored. This quadratic term is handled with a fraction with a constant plus a constant times x in the numerator.

$$\frac{5x^3 - 3x^2 + 2x - 1}{x^2(x^2 + 1)} = \frac{A}{x} + \frac{B}{x^2} + \frac{Cx + D}{x^2 + 1}$$

Multiply by the common denominator and compare coefficients.

$$5x^3 - 3x^2 + 2x - 1 = Ax(x^2 + 1) + B(x^2 + 1) + (Cx + D)x^2$$

$$5x^3 - 3x^2 + 2x - 1 = (A + C)x^3 = (B + D)x^2 + Ax + B$$

$B = -1, D = -2, A = 2$ and $C = 3$ so the integral is written as

$$\int \frac{5x^3 - 3x^2 + 2x - 1}{x^4 + x^2} \, dx = 2\int \frac{dx}{x} - \int \frac{dx}{x^2} + \int \frac{3x - 2}{x^2 + 1} \, dx$$

$$\int \frac{5x^3 - 3x^2 + 2x - 1}{x^4 + x^2} \, dx = 2\ln|x| + x^{-1} + \frac{3}{2}\ln(x^2 + 1) - 2\tan^{-1}x + K$$

That last integral came from a Table of Integrals. Quadratic partial fractions usually produce very difficult integrals.

10-7   Integrals from Tables

One of the best techniques of integration is to use the table of integrals found in most texts. A table of integrals is found in the Mathematical Tables included at the back of this book.

Some instructors do not allow the use of tables on tests. We do not share that view. Why take up precious memory space with formulas that are available in an inexpensive mathematical table? Regardless of your instructor, you will eventually want to use tables, and these examples will give you an introduction to the process. Most tables are organized by categories: trigonometric, logarithmic, exponential, quadratics or fractions, or whatever. The examples we have chosen are, hopefully, appropriate for what you will encounter.

My sister Liebie never uses tables. She likes to show off and work all the integrals.

Example 10-25 Find $\int \dfrac{dx}{x\sqrt{x^2 + 4}}$.

Solution: An integral in this form is in the tables. It reads

$$\int \frac{du}{u\sqrt{a^2 + u^2}} = -\frac{1}{a}\ln\left|\frac{\sqrt{a^2 + u^2} + a}{u}\right|$$

Make the identification that $x = u$ and $a = 2$ and write down the integral.

$$\int \frac{dx}{x\sqrt{x^2 + 4}} = -\frac{1}{2}\ln\left|\frac{\sqrt{4 + x^2} + 2}{x}\right|$$

Example 10-26 Find $\int \dfrac{dx}{x(1 + 2x)}$.

Solution: An integral in this form is in the tables. It reads

$$\int \frac{du}{u(a + bu)} = \frac{1}{a}\ln\left|\frac{u}{a + bu}\right|$$

Make the identification that $x = u$, $a = 1$, and $b = 2$ and write down the integral.

$$\int \frac{dx}{x(1 + 2x)} = \frac{1}{1}\ln\left|\frac{x}{1 + 2x}\right| = \ln\left|\frac{x}{1 + 2x}\right|$$

Using the tables is this easy. Go slowly and make sure you are identified with the correct integral and make the substitutions. Remember that some tables include the constant of integration and some do not. If you are working an indefinite integral be sure to include the constant in any calculation.

10-8   Approximate Methods

When all else fails, use numeric integration! For the definite integral

$$y = \int_a^b (\text{some impossible to integrate function of } x)dx$$

the area under the curve of y vs. x from a to b is the value of the integral.

There are several different approximation methods. The general approach to numerical integration will be illustrated by a relatively simple one, the trapezoidal rule or method. As the name implies the area to be determined is divided up into trapezoids. Consider the area under some general curve as shown in Fig. 10-12. Divide the region within the limits into several narrow regions bounded by the vertical lines at $x_0, x_1, x_2 \ldots$ with a fixed width Δx between each line. Corresponding to each of the $x_0, x_1, x_2 \ldots$ values is a value of the function $f_0, f_1, f_2, \ldots$. The first two regions are shown in exploded view and better illustrate that the curve is approximated by a straight line creating a collection of trapezoids.

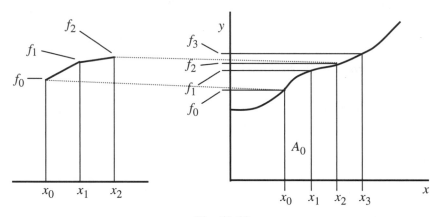

Fig. 10-12

The area of the first region is the large rectangle with dimensions Δx and f_0 plus the small rectangle on top of it with base Δx and height $f_1 - f_0$.

The area of the first region is

$$A_0 = f_0 \Delta x + \frac{1}{2} [f_1 - f_0] \Delta x$$

The area of the second region is

$$A_1 = f_1 \Delta x + \frac{1}{2} [f_2 - f_1] \Delta x$$

By analogy the next region is

$$A_2 = f_2 \Delta x + \frac{1}{2} [f_3 - f_2] \Delta x$$

The total area taken over all the intervals is the sum of these individual areas.

$$A = A_0 + A_1 + A_2 + \cdots$$

$$A = f_0 \Delta x + \frac{1}{2}[f_1 - f_0]\Delta x + f_1 \Delta x + \frac{1}{2}[f_2 - f_1]\Delta x + f_2 \Delta x$$

$$+ \frac{1}{2}[f_3 - f_2]\Delta x + \cdots$$

Multiplying and collecting terms, we calculate

$$A = f_0 \Delta x - f_0 \frac{\Delta x}{2} + f_1 \Delta x + f_1 \frac{\Delta x}{2} - f_1 \frac{\Delta x}{2} + f_2 \Delta x + f_2 \frac{\Delta x}{2} - f_2 \frac{\Delta x}{2} + \cdots$$

Certain of these terms combine and the pattern that emerges is

$$A = f_0 \frac{\Delta x}{2} + f_1 \Delta x + f_2 \Delta x + \cdots$$

Continuing the pattern, the last area, call it n, has an associated term $f_n \frac{\Delta x}{2}$.

Another way of writing the area sum is

$$A = \frac{\Delta x}{2}[f_0 + 2f_1 + 2f_2 + \cdots + 2f_{n-1} + f_n]$$

The f_0 term is the left-most limit and the f_n term is the right-most limit. The width of each individual region, Δx, is the extent of the limits $(b - a)$ divided by n, the number of intervals.

Apply this technique to a simple and then a not so simple problem.

Example 10-27   Find the value of the definite integral of the curve $y^2 = x^{2.9}$ from $x = 0$ to $x = 2$ using the trapezoidal rule for the area under the curve.

Solution: Take the square root of both sides of the equation to find y as a function of x : $y = x^{1.7}$. This curve is something less than a quadratic. The integral to be evaluated is $\int_0^2 x^{1.7}dx$ and it is the area under $y = x^{1.7}$ from $x = 0$ to $x = 2$.

A rough sketch of the curve is shown in Fig. 10-13.

Use 10 intervals so that $\Delta x = 0.20$ and $\Delta x/2 = 0.10$.

Following the trapezoidal rule for area we write

$$A = \frac{\Delta x}{2}[f_0 + 2f_{0.2} + 2f_{0.4} + \cdots + 2f_{1.8} + f_{2.0}]$$

Adding the numbers

$$A = 0.10[0 + 0.13 + 0.42 + 0.84 + 1.37 + 2 + 2.73 + 3.54$$
$$+ 4.45 + 5.43 + 3.25] = 2.42$$

Check this answer by performing the integral and evaluating $\int_0^2 x^{1.7}dx$.

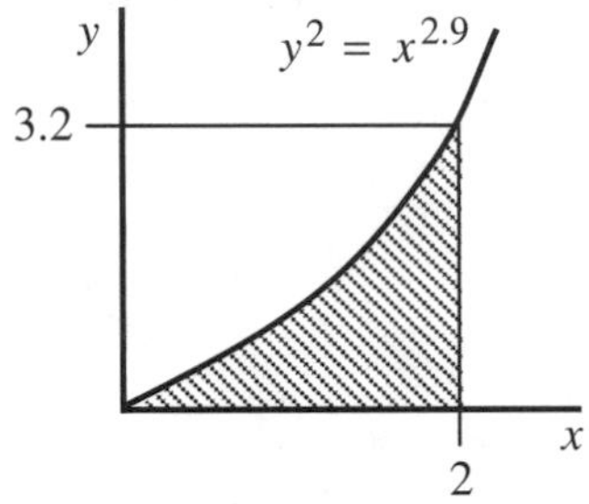

Example 10-28 Evaluate $\int_0^1 \sqrt{1 - x^5}dx$ using the trapezoidal rule with 4 intervals.

Solution: In the range between $x = 0$ and $x = 1$, the function $\sqrt{1 - x^5}$ goes from 1 to 0. A detailed curve is not necessary to the calculation. However, a rough sketch is shown in Fig. 10-14. Four intervals means that $\Delta x = 0.25$ and using the formula for the trapezoidal rule

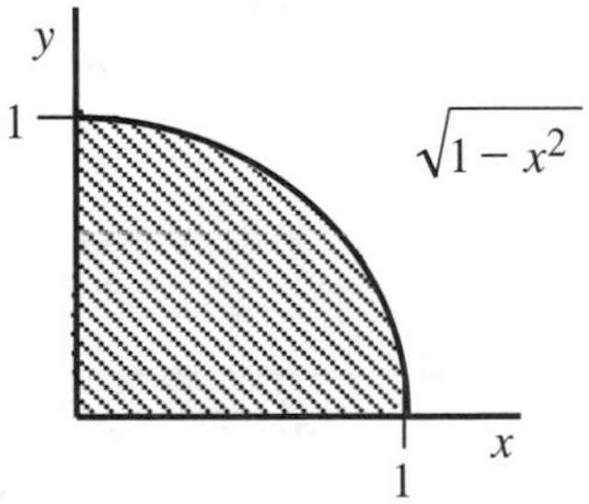

$$A = \frac{\Delta x}{2} [f_0 + 2f_{0.25} + 2f_{0.50} + 2f_{0.75} + f_1]$$

$$A = \frac{0.25}{2} [1 + 2.00 + 1.97 + 1.75 + 0] = 0.84$$

Fig. 10-13

Fig. 10-14

Example 10-29 Suppose that $2000 is invested in a fund at the beginning of each of 10 years and that the average rate of return is 20% per year. What is the total value of this fund at the end of the 10 years?

Solution: Visualize the process with the aid of the time line. The first $2000 grows compounded at 20% for 10 years so this is $2000e^{0.20(10)}$.

Refer to Example 9-7 for a discussion of the effective rate for continuous compounding.

Use the continuous compounding as an approximation, because the $2000 is deposited at the beginning of each interval rather than in small increments throughout the interval.

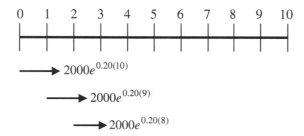

The first $2000 grows compounded at 20% for 10 years to an amount $2000e^{0.20(10)}$.

The second $2000 grows compounded at 20% for 9 years to an amount $2000e^{0.20(9)}$.

The third $2000 grows compounded at 20% for 8 years to an amount $2000e^{0.20(8)}$.

And so on through the 10 deposits.

The general expression for the terms is $2000e^{0.20(10-t)}$, where t goes from 0 to 9. The total amount at the end of the 10 years is $2000 times the 10 years plus the interest earned on the different intervals. This can be expressed as a sum

$$\sum_{n=0}^{n=10} 2000e^{0.20(10-t_n)}\Delta t \quad \text{with} \quad t_0 = 0, t_1, \ldots$$

This $\sum_{n=0}^{n=10}$ is the notation for adding, or summing, all the exponents. If the funds were deposited continuously and the compounding was continuous, then this sum would be an integral. Many programs for placing funds in a compounding account are monthly throughout the year rather than once at the beginning of the year, making those programs closer to the continuous model. For this case the integral is an approximation! In mathematical language, the funds being deposited continuously means that the interval goes to zero, or n, the number of intervals, goes to infinity. In symbolic language we write

$$\lim_{n\to\infty}\sum_{n}2000e^{0.20(10-t_n)}\Delta t = \int_0^{10}2000e^{0.20(10-t)}dt$$

The exponent can be reworked to $e^{0.20(10-t)} = e^2e^{-0.20t}$ so the integral for the total amount is

$$A = 2000e^2\int_0^{10}e^{-0.20t}dt$$

Make a change of variable and integrate.

$$A = \frac{2000}{-0.20}e^2\int_0^{10}e^{-0.20t}d(-0.20t) = -10{,}000\,e^2\Big[e^{-0.20t}\Big]_0^{10}$$

$$A = -10{,}000e^2\,[e^{-2} - 1] = 10{,}000e^2\Big[1 - \frac{1}{e^2}\Big] = \$63900$$

This number is lower than the actual amount if the funds were deposited at the beginning of each year. If the funds were placed continuously throughout the year then this number is correct. Programs for making calculations similar to this one are usually found in financial calculators. If you have one, check this answer with the answer from your calculator. Your calculator uses a calculating algorithm similar, if not identical, to the summation notation used earlier in the problem.

It's a Wrap

✔ Find volumes by discs and shells

✔ Know the formula for arc length

✔ Wrap the arc length for surfaces of revolution

✔ Integrate by changing variables

✔ Use integration by parts

✔ Apply partial fractions

✔ Know how to use tables

✔ Be able to apply approximation methods

Test Yourself

PROBLEMS

1. Find the volume of the cone generated by rotating the line $y = 3x$ about the y-axis and bounded by the line $y = 4$. Use the method of disks.
2. Find the volume generated by rotating the area between $y = x^2 - x^3$ and the x-axis from $x = 0$ to $x = 1$ about the y-axis.
3. Solve the integral $\int \dfrac{4x - 1}{x^2 + x - 2}\, dx$ using partial fractions.
4. Integrate $\int x \sin x\, dx$ by parts.
5. Sales of a certain product follow a cyclic pattern superimposed on constant sales in the form $S = 25 + 10\cos(\pi t/8)$ with S measured in hundreds of items sold and t is the month measured from June 1st each year. Find the average monthly sales during the summer months of June, July, and August.
6. Use the trapezoidal rule with 10 intervals to find the area under the curve $y = \sqrt{x}$ from $x = 0$ to $x = 1$.

ANSWERS

1. Draw a diagram including the disk. The volume of the disk is $\pi x^2 dy$ so the total volume is the integral

$$V = \int_0^4 \pi x^2 dy = \int_0^4 \frac{\pi y^2}{9}\, dy = \frac{\pi}{27} y^3 \Big|_0^4 = \frac{\pi(4)^3}{27} = 7.4$$

2. Sketch the curve. For small x the x^2 term dominates, and for larger x the x^3 term begins to dominate. In this problem it would be very difficult to use disks. First we would have to find the equation for $x =$. This is no small task. Next we would have to find the peak in the curve and subtract the portion of the curve up to this peak from the portion of the curve from this peak to 1. In this problem the best procedure is to use what some authors call *the method of shells*, where the little rectangle shown as dx is rotated about the

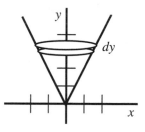

y-axis with the integral carried out over 0 to 1 in x. The shell has height $y = x^2 - x^3$, width dx, and is rotated ($2\pi x$) about the y-axis. The volume integral is

$$V = \int_0^1 2\pi x(x^2 - x^3)dx = 2\pi \left[\frac{x^4}{4} - \frac{x^5}{5} \right] \Big|_0^1 = 2\pi \left[\frac{1}{4} - \frac{1}{5} \right] = 0.31$$

3. The denominator can be factored so write the partial fraction form.

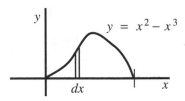

$$\frac{4x - 1}{x^2 + x - 2} = \frac{A}{(x + 2)} + \frac{B}{(x - 1)}$$

Multiply by the common denominator and compare coefficients.

$$4x - 1 = A(x - 1) + B(x + 2) = (A + B)x - A + 2B$$

$$A + B = 4 \quad 3B = 3 \quad \Rightarrow \quad B = 1$$
$$-A + 2B = -1 \quad A = 3$$

Write the integral in partial fraction form.

$$\int \frac{4x - 1}{x^2 + x - 2} dx = \int \frac{3}{(x + 2)} dx + \int \frac{1}{(x - 1)} dx = 3 \ln|x + 2|$$

$$+ \ln|x - 1| + K$$

4. Using the form $\int u dv = uv - \int v du$ and let $u = x$ so $du = dx$ and $dv = \sin x \, dx$ so $v = -\cos x$. The integration is then

$$\int x \sin x dx = -x \cos x + \int \cos x dx = -x \cos x + \sin x$$

5. Find the average sales with the average value integral.

$$Avg = \frac{1}{3 - 0} \int_0^3 25 dt + \int_0^3 10 \cos \frac{\pi t}{8} dt = \frac{1}{3} \left[25t \Big|_0^3 + \frac{80}{\pi} \int_0^3 \cos \frac{\pi t}{8} d \frac{\pi t}{8} \right]$$

$$Avg = \frac{1}{3} \left[75 + \frac{80}{\pi} \sin \frac{\pi t}{8} \Big|_0^3 \right] = \frac{1}{3} [75 + 23.5] = 32.8$$

6. Use the trapezoidal formula with $\Delta x = 0.10$.

$$A = \frac{\Delta x}{2} [f_0 + 2f_1 + 2f_2 + \cdots + 2f_{n-1} + f_n]$$

$$A = \frac{0.1}{2} [0 + 0.316 + 0.447 + 0.548 + 0.632 + 0.707$$

$$+ 0.775 + 0.837 + 0.894 + 0.949 + 1] = 0.355$$

MATHEMATICAL TABLES

Geometry

r is radius, h is height, a and b are sides

	Perimeter	Area	Volume
square side a	$4a$	a^2	
rectangle sides a and b	$2a + 2b$	ab	
circle radius r	$2\pi r$	πr^2	
sphere radius r		$4\pi r^2$	$(4/3)\pi r^3$
cylinder r and h		$2\pi rh$	$\pi r^2 h$
cone r and h		$\pi r\sqrt{r^2 + h^2}$	$(\pi/3)r^2 h$
trapezoid a, b, h		$(1/2)(a + b)h$	
triangle b and h		$(1/2)bh$	

Algebra

Any quadratic equation of the form $ax^2 + bx + c = 0$ has solution $x = \dfrac{-b \pm \sqrt{b^2 - 4ac}}{2a}$.

Factorials $0! = 1$, $1! = 1$, $2! = 2 \cdot 1! = 2 \cdot 1$, $3! = 3 \cdot 2! = 3 \cdot 2 \cdot 1$, etc.

Binomial expansion $(a + b)^n = \dfrac{a^n}{0!} + \dfrac{na^{n-1}b}{1!} + \dfrac{n(n - 1)a^{n-2}b^2}{2!} + \ldots$

for $b^2 < a^2$.

Conics

parabola $\quad y = ax^2 + bx + c$

circle $\quad x^2 + y^2 = r^2$

ellipse $\quad ax^2 + by^2 = c^2 \quad$ or $\quad \dfrac{x^2}{a^2} + \dfrac{y^2}{b^2} = 1$

hyperbola $\quad \pm ax^2 \mp by^2 = c^2 \quad$ or $\quad \pm\dfrac{x^2}{a^2} \mp \dfrac{y^2}{b^2} = 1$

Trigonometry

$$\sin\theta = \frac{\text{Opposite}}{\text{Hypotenuse}} = \frac{b}{c}$$

$$\cos\theta = \frac{\text{Adjacent}}{\text{Hypotenuse}} = \frac{a}{c}$$

$$\tan\theta = \frac{\text{Opposite}}{\text{Adjacent}} = \frac{b}{a}$$

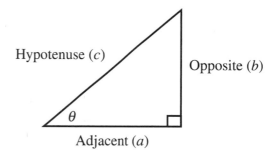

$$\text{cosec}\,\theta = \frac{1}{\sin\theta} \qquad \sec\theta = \frac{1}{\cos\theta} \qquad \cot\theta = \frac{1}{\tan\theta}$$

Law of sines $\dfrac{a}{\sin A} = \dfrac{b}{\sin B} = \dfrac{c}{\sin C}$

Law of cosines $c^2 = a^2 + b^2 - 2ab\cos C$

$360° = 2\pi\,\text{radians}$

Trigonometric Functions

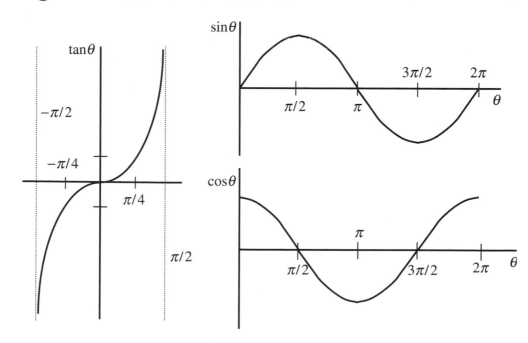

Trigonometric Identities

$$\sin^2\theta + \cos^2\theta = 1 \qquad a^2 + b^2 = c^2$$

$$\sin\theta = \cos(90° - \theta) \qquad \cos\theta = \sin(90° - \theta) \qquad \tan\theta = \cot(90° - \theta)$$

$$\sin(\alpha \pm \beta) = \sin\alpha\cos\beta \pm \cos\alpha\sin\beta$$

$$\cos(\alpha \pm \beta) = \cos\alpha\cos\beta \mp \sin\alpha\sin\beta$$

$$\sin 2\alpha = 2\sin\alpha\cos\alpha \qquad \cos 2\alpha = \cos^2\alpha - \sin^2\alpha$$

$$\tan(\alpha \pm \beta) = \frac{\tan\alpha \pm \tan\beta}{1 \mp \tan\alpha\tan\beta}$$

Exponents and Logarithms

$$a^m \cdot a^n = a^{m+n} \qquad \frac{a^m}{a^n} = a^{m-n} \qquad (a^m)^n = a^{m\cdot n} \qquad a^{-m} = \frac{1}{a^m}$$

$$\log_a u + \log_a v = \log_a(uv) \qquad \log_a u - \log_a v = \log_a\frac{u}{v} \qquad \log_a u^n + n\log_a u$$

Differential and Integral Formulas

$$d(ax) = adx \qquad\qquad \int adx = ax$$

$$d(u + v) = du + dv \qquad\qquad \int (du + dv) = u + v$$

$$d(uv) = udv + vdu \qquad\qquad \int udv = u\int dv - \int vdu = uv - \int vdu$$

$$d\frac{u}{v} = \frac{vdu - udv}{v^2} \qquad\qquad \int u\frac{dv}{dx}dx = uv - \int v\frac{du}{dx}dx$$

$$dx^n = nx^{n-1}dx \qquad\qquad \int x^n dx = \frac{x^{n+1}}{n+1}$$

$$de^x = e^x dx \qquad\qquad \int e^x dx = e^x$$

$$d\ln x = \frac{1}{x}dx \qquad\qquad \int \ln x dx = x\ln x - x$$

$$d \sin x = \cos x dx \qquad \int \sin x dx = -\cos$$

$$d \cos x = -\sin x dx \qquad \int \cos x dx = \sin x$$

$$d \tan x = \sec^2 x dx \qquad \int \tan x dx = -\ln(\cos x)$$

$$d \sec x = \tan x \sec x dx \qquad \int \sec x dx = \ln(\sec x + \tan x)$$

$$d \csc x = -\cot x \csc x dx \qquad \int \csc x dx = \ln(\csc x - \cot x)$$

$$d \cot x = -\csc^2 x dx \qquad \int \cot x dx = \ln(\sin x)$$

Integral Formulas

$$\int \frac{dx}{\sin x} = \ln(\csc x - \cot x)$$

$$\int \frac{dx}{\cos x} = \ln(\sec x + \tan x)$$

$$\int x \sin x dx = \sin x - x \cos x$$

$$\int x \cos x dx = \cos x + x \sin x$$

$$\int x \ln x dx = \frac{x^2}{2} \ln x - \frac{x^2}{4}$$

$$\int e^{ax} dx = \frac{e^{ax}}{a}$$

$$\int x e^{ax} dx = \frac{e^{ax}}{a^2}(ax - 1)$$

$$\int \frac{dx}{x^2 + a^2} = \frac{1}{a} \tan^{-1} \frac{x}{a}$$

$$\int \frac{dx}{a^2 - x^2} = \frac{1}{2a} \ln \frac{a + x}{a - x}$$

$$\int \frac{dx}{x^2 - a^2} = \frac{1}{2a} \ln \frac{x - a}{x + a}$$

$$\int (a + bx)^n \, dx = \frac{(a + bx)^{n+1}}{(n + 1)b} \text{ except } n = -1$$

$$\int \frac{dx}{a + bx} = \frac{1}{b} \ln (a + bx)$$

$$\int \frac{xdx}{a + bx} = \frac{1}{b^2} [a + bx - a \ln (a + bx)]$$

$$\int \frac{dx}{a + bx^2} = \frac{1}{\sqrt{ab}} \tan^{-1} \frac{x\sqrt{ab}}{a}$$

$$\int \frac{xdx}{a + bx^2} = \frac{1}{2b} \ln \left(x^2 + \frac{a}{b} \right)$$

$$\int \sqrt{a + bx} \, dx = \frac{2}{3b} \sqrt{(a + bx)^3}$$

$$\int x\sqrt{a + bx} \, dx = -\frac{2(2a - 3bx)\sqrt{(a + bx)^3}}{15b^2}$$

$$\int \sqrt{x^2 \pm a^2} \, dx = \frac{1}{2} \left[x\sqrt{x^2 \pm a^2} \pm a^2 \ln (x + \sqrt{x^2 \pm a^2}) \right]$$

$$\int \frac{dx}{\sqrt{x^2 \pm a^2}} = \ln \left(x + \sqrt{x^2 \pm a^2} \right)$$

$$\int \sqrt{a^2 - x^2} \, dx = \frac{1}{2} \left[x\sqrt{a^2 - x^2} + a^2 \sin^{-1} \left(\frac{x}{a} \right) \right]$$

$$\int \frac{dx}{\sqrt{a^2 - x^2}} = \sin^{-1} \left(\frac{x}{a} \right)$$

Index